AGRICULTURAL AND ENVIRONMENTAL POLICIES

OPPORTUNITIES FOR INTEGRATION

ORGANISATION FOR ECONOMIC CO-OPERATION AND DEVELOPMENT

Pursuant to article 1 of the Convention signed in Paris on 14th December 1960, and which came into force on 30th September 1961, the Organisation for Economic Co-operation and Development (OECD) shall promote policies designed:

- to achieve the highest sustainable economic growth and employment and a rising standard of living in Member countries, while maintaining financial stability, and thus to contribute to the development of the world economy;
- to contribute to sound economic expansion in Member as well as non-member countries in the process of economic development; and
- to contribute to the expansion of world trade on a multilateral, non-discriminatory basis in accordance with international obligations.

The original Member countries of the OECD are Austria, Belgium, Canada, Denmark, France, the Federal Republic of Germany, Greece, Iceland, Ireland, Italy, Luxembourg, the Netherlands, Norway, Portugal, Spain, Sweden, Switzerland, Turkey, the United Kingdom and the United States. The following countries acceded subsequently through accession at the dates indicated hereafter: Japan (28th April 1964), Finland (28th January 1969), Australia (7th June 1971) and New Zealand (29th May 1973).

The Socialist Federal Republic of Yugoslavia takes part in some of the work of the OECD (agreement of 28th October 1961).

Publié en français sous le titre:

POLITIQUES
DE L'AGRICULTURE ET DE
L'ENVIRONNEMENT
POSSIBILITÉS D'INTÉGRATION

PREFACE

Agricultural activities, as compared to industrial activities, have been in harmony over a long period with the environmental aspirations of OECD societies. However, with modern technology and increased support for agriculture, conflicts have arisen between agricultural and environmental objectives.

OECD governments have recognised the potential damage that agriculture can inflict on the environment and measures are taken now to minimise such damage, pre-empt conflicts and to enhance the positive impacts. This report presents the various policy alternatives to harmonize agricultural and environmental objectives that are available now and likely to emerge in the near future.

OECD governments have now reached a consensus that in future they will improve the integration of their agricultural and environmental policies to exploit the opportunities available and to realize policy objectives in both areas with the minimum of economic and environmental cost.

The main body of this report examines these problems and presents the policy alternatives with particular emphasis on problems arising from the use of agricultural chemicals, the surplus of animal manure, soil erosion and changes to the agricultural landscape. Annex I of the report describes the way countries in the recent past approached the integration of agricultural and environmental policies. Annex II presents a synopsis of case studies carried out in selected Member countries concerning the major agricultural and environmental issues addressed in the report. The two annexes have been derestricted under the responsibility of the Secretary-General and do not necessarily represent the views of the governments of OECD Member countries.

Also available

PRICING OF WATER SERVICES (1987)
(97 87 02 1) ISBN 92-64-12921-9   146 pages   £8.00   US$17.00   FF80.00   DM38.00

OECD ENVIRONMENTAL DATA/DONNÉES OCDE SUR L'ENVIRONNEMENT – COMPENDIUM 1987. Bilingual
(97 87 05 3) ISBN 92-64-02960-5   366 pages   £20.00   US$42.00   FF200.00 DM86.00

WATER POLLUTION BY FERTILIZERS AND PESTICIDES (1986)
(97 86 02 1) ISBN 92-64-12856-5   144 pages   £6.00   US$12.00   FF60.00   DM27.00

RURAL PUBLIC MANAGEMENT (1986)
(42 86 02 1) ISBN 92-64-12858-1   86 pages   £5.00   US$10.00   FF50.00   DM25.00

Prices charged at the OECD Bookshop.

*THE OECD CATALOGUE OF PUBLICATIONS and supplements will be sent free of charge
on request addressed either to OECD Publications Service,
2, rue André-Pascal, 75775 PARIS CEDEX 16, or to the OECD Distributor in your country.*

TABLE OF CONTENTS

With modern technology and increased support for agriculture, a conflict has arisen between agricultural and environmental objectives. This is, in part, due to a failure to recognise the interdependence of agriculture and environment. The intensification of agricultural practices, stimulated by agricultural policy, has had an adverse effect on the environment. At the same time it must also be realised that agriculture can and does make a significant positive contribution to the quality of the environment. The impact of pollution from other sources on agriculture is also an issue of concern in some countries.

The essential theme of this report is the need for the integration of policies so that whenever possible mutual benefits are realised and, whenever necessary, conscious trade-offs are made between competing agricultural and environmental objectives.

An integrated approach requires environmental considerations to be taken fully into account at an early stage in the development and implementation of agricultural policies. Similarly, during the formulation and implementation of environmental policies full consideration must be given to the potential impacts on agricultural production, incomes and prices.

In developing agricultural and related development policies, consideration needs to be given to a trilogy of interdependent factors: i) the need to enhance the positive contribution which agriculture can make to the environment; ii) the need to reduce agricultural pollution; and iii) the importance of adapting agricultural policies so that they take full account of the environment.

-- First, the positive role of agriculture which does not cause pollution or environmental damage can be enhanced through the introduction of management agreements and other similar arrangements which if, of sufficient duration, will produce the expected environmental benefits can be realised. In certain circumstances farmers may need to be compensated for the lost net value of production and additional maintenance costs. The main use of these agreements and arrangements should be to improve landscape amenity and conservation values.

-- Second, in order to reduce agricultural pollution, different possible measures need to be considered, either individually or in combination. In some cases, the setting and enforcement of standards will be most efficient. In other cases, the implementation of advisory procedures, or the application of economic measures such as incentives or charges may be superior to regulatory enforcement. In all cases the Polluter-Pays Principle should be observed. Efforts should be made to overcome the perceived difficulties associated with applying this principle to the control of agricultural pollution from diffuse sources.

-- Third, policies principally designed to achieve agricultural and other sector objectives, such as the reduction of agricultural support, can and should in addition be targeted to produce maximum environmental benefits.

For all three factors integration is likely to be more effective when technical support is given to farmers and when research, development and monitoring programmes pursue the integration of agricultural and environmental policies. With regard to the third factor it is recommended that support reduction programmes be targeted so that they simultaneously reduce surpluses, reduce agricultural pollution and enhance environmental quality. Targeting mechanisms include the provision of financial incentives for the set-aside of land which is a source of pollution and the restriction of the programmes to areas which contribute to environmental problems.

Environmental policies should, recognising that integration is a two-way process, be formulated in a way which facilitates the development of an efficient and prosperous agricultural sector. For example, in formulating air emission standards full consideration should be given to the effects of industrial pollution regulations on agricultural resource values. More research on the effects of pollution and environmental regulation on agriculture and their economic consequences would enhance decision-making in this area.

ADMINISTRATIVE INTEGRATION

Successful integration requires policy-makers to give full consideration to, and accept responsibility for, the effects of their policies on the objectives of all other sectors. This is as true for the effects of environmental policies on agricultural policies as it is for the effects of agricultural policies on the environment.

One necessary pre-condition to this concept of "responsibility" is that policy-makers must jointly agree to a set of objectives for the interacting and interdependent sectors. Greater progress will be made when these objectives are clear, measurable and time specific. Progress will also be more likely when there is a publicly made commitment, at Ministerial level, to the integration of environmental and agricultural policies and when administrative structures are created to promote integration.

One way of achieving more effective integration is to change institutional arrangements so that there is a creative dialogue between those responsible for protecting the environment and those responsible for agriculture. Another is to establish administrative procedures which require collaboration among agencies such as the mandatory referral of policy proposals to other departments. The application of environmental impact assessment procedures to agricultural activities and policies is another way of improving administrative integration.

Because of the regional disparities in physical, social and economic conditions, successful integration will usually require the development of strategies which rely upon a mixture of policies. Some of these social and economic strategies will be national in character while others, particularly in larger countries, will be regional. Experience in some of these larger countries suggests that the opportunities for integration are often greater at a regional level.

OPPORTUNITIES FOR SUCCESSFUL POLICY INTEGRATION

There are many existing opportunities for the integration of agricultural and environmental policies and others are emerging as agricultural policies change: the distinction between existing and emerging opportunities, however, is somewhat arbitrary. Some of the most significant opportunities and three interdependent factors to be considered in achieving successful integration are identified in Figure 1.

Existing opportunities

Existing opportunities for the better integration of agricultural and environmental policies include:

-- developing research and advisory programmes with a view to giving greater emphasis and broader consideration to environmental objectives;

-- strengthening the provision of education and advisory services with a view to improving the use of agricultural inputs and modifying agricultural practices to minimise damage to the environment;

-- encouraging or requiring farmers to prepare management plans which indicate how they will use inputs and adopt practices so as to protect and enhance the environment;

-- entering into management agreements and other arrangements with farmers to improve landscape amenity and nature conservation values;

-- removing impediments to the adoption of environmentally favourable practices such as integrated pest management practices by farmers who wish to remain eligible for the receipt of financial assistance;

-- introducing charges on inputs such as fertilizers and pesticides, and, also, on practices such as the spreading of animal manure to contribute to the cost of advisory, research and other activities designed to prevent and control agricultural pollution;

-- enforcing existing regulations more stringently;

-- further harmonising the standards and procedures used in regulating the use of agricultural inputs, including food quality and labelling standards, changing product quality standards and influencing consumer preferences so that they reduce incentives to use pesticides; and

-- making income, capital and land taxation policies neutral to agricultural and environmental objectives.

It is noted that many countries have already begun to recognise many of these existing opportunities for integration. It would appear, however, that there is still considerable room for progress. In particular, policies which further develop the role of farmers in maintaining and improving landscape amenity and nature conservation objectives are being advocated. At the same

Figure 1

OPPORTUNITIES FOR THE SUCCESSFUL INTEGRATION OF
AGRICULTURAL AND ENVIRONMENTAL POLICIES

A D M I N I S T R A T I V E   I N T E G R A T I O N

-- AGREED ENVIRONMENTAL
   QUALITY OBJECTIVES

-- RECOGNISED RESPONSIBILITY

-- REGIONAL APPROACH

P O L I C Y   I N T E G R A T I O N

EMERGING OPPORTUNITIES   EXISTING OPPORTUNITIES

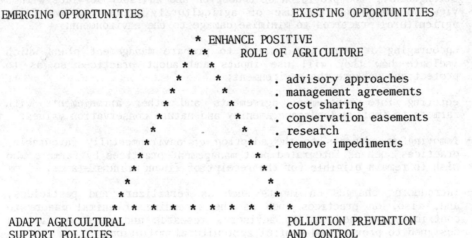

```
              * ENHANCE POSITIVE
           * *       ROLE OF AGRICULTURE
          *    *    *
         *      *  . advisory approaches
        *        * . management agreements
       *          *. cost sharing
      *            *. conservation easements
     *             *. research
    *              *. remove impediments
   *              *
  *              *
 * * * * * * * * * *
```

ADAPT AGRICULTURAL   POLLUTION PREVENTION
SUPPORT POLICIES   AND CONTROL

. production incentives   . advisory approaches
. income support   . standards
. set-aside   . charges & levies
. cross-compliance   . enforcement
. diversification   . research & monitoring
. quotas   . remove impediments
. structural adjustment

time, in implementing pollution prevention and control measures it is suggested that more account needs to be taken of the effects of environmental policies on agriculture in general and agricultural incomes in particular.

There is a need, particularly in areas where there is significant agricultural pollution, for a more precise definition of unacceptable agricultural practices which cause pollution. A major problem in many countries is the degree to which existing land-use regulations are poorly defined and, even where well defined, not enforced.

There is strong pressure, spread over many if not all Member countries and, indeed, worldwide, for prices to reflect the true cost to society of the provision of goods and services. In the context of agriculture and the environment this means that farmers should pay for the full marginal cost of any natural resources used, particularly those which are in scarce supply such as water. This concept applies to all industries and is just as relevant to agriculture as it is to the users of all other resources.

Pollution can be reduced and biological diversity on farmland can be increased if farmers are encouraged to use farm manure, mineral fertilizers and pesticides more sparingly and more precisely. This can be achieved with new technological advances backed up by sound advice and appropriate management guides.

There are also a large number of opportunities to improve the environment at modest cost and little inconvenience to the farmer. Leaving field margins unfertilized, leaving aside small pockets of land, planting trees and retaining hedges can significantly improve the environment.

Emerging opportunities

Governments of OECD Member countries have agreed on principles to reform agricultural policies and in general to reduce agricultural supports and thereby bring about a more efficient allocation of resources to the benefit of consumers and society as a whole. Environmental concerns over the intensity of agricultural production provide added impetus to efforts to reduce supports to agriculture. The reform process provides opportunities for the integration of agricultural and environmental policies which will be mutually beneficial to the achievement of agricultural reform and environmental objectives. As measures to reduce production and, in particular, to limit price supports are realised, major opportunities to achieve these objectives and simultaneously improve the environment can be expected to emerge.

The reduction of output related support measures or the introduction of quantitative restrictions on production might result in decreased agricultural pollution, particularly in areas where the intensity of agricultural production is high.

As supports to agriculture are reduced, in areas where landscape amenity, cultural heritage, wildlife habitat and species diversity values are important, the provision and redirection of specific financial assistance to encourage farmers to continue with environmentally-favourable practices may be necessary. Examples of such practices include the repair of stone walls and terraces, the maintenance of historic farm buildings and the leaving of field margins unsprayed.

Other changes to agricultural policy which may occur in the near future and thus offer opportunities for the development of integrated policies, include changes to tariff arrangements, the introduction of cross-compliance requirements, direct income support to farmers, the establishment of land set-aside programmes and the introduction of quotas on inputs and outputs.

One opportunity for achieving the improved integration of agricultural and environmental policies is the attachment of cross-compliance provisions to price support. This recommended approach requires farmers to comply with a set of pre-specified land-use conditions and practices in return for the receipt of government payments. Cross-compliance can only be introduced when farmers receive direct support such as deficiency payments.

Cross-compliance is also being used to reduce the tendency of some production incentive policies to expand cultivation into wetlands and other environmentally fragile areas. One example of this is the refusal of financial assistance to farmers who begin to crop highly erodible land without an approved conservation plan or who drain and begin to crop wetlands.

Although land set-aside policies have a number of disadvantages and only offer a second best interim solution, they have the potential to achieve simultaneously agricultural, budgetary and environmental goals. Opportunities for realising these benefits will be greatest when:

-- set-aside programmes are targeted to areas with environmental problems, including key production areas;

-- the time period for set-aside is sufficient to induce farmers to diversify into environmentally favourable activities such as certain types of forestry;

-- the set-aside land is selected under a tender or similarly based system which includes an assessment of the likely off-site impacts from continued production and the potential budget savings from setting the land aside;

-- the programmes exclude short term rotational set-aside; and

-- the set-aside programme is presented so as to encourage farmers to enhance their role in conserving, maintaining and improving the environment.

Opportunities to set aside highly productive land in filter strips along the edge of streams and in groundwater areas should not be discounted. In some situations it is noted that such an approach can produce substantial environmental benefits and budgetary savings from reduced surplus production.

1. AGRICULTURE AND THE ENVIRONMENT

Agriculture in its historical and national context: Harmony and Conflict

Agriculture has had a long positive association with the environment. All OECD countries have sought to promote agricultural development by funding research, by providing sophisticated extension and advisory services, by giving other forms of assistance and by stimulating production through the provision of subsidies. As a result agricultural production has quadrupled in this century alone, contributing to accelerated urban development, industrial growth and expansion of the service sector. But at the same time agricultural pollution has increased and the quality of a number of rural landscapes has declined.

Farming has become much more mechanised and more intensive, with greater regional and on-farm specialisation and greater regional concentration. Exploiting mechanisation and technology, replacing man and beast with energy from fossil fuels, strengthening the productivity of the soil and crop yields with fertilizers and pesticides*, agriculture has evolved to a state where short term profits can be made without maintaining the traditional harmony and interdependence between agriculture and the environment which has existed for centuries.

While agriculture still makes a significant contribution to the landscape in many areas, because of a failure to integrate agricultural and environmental policies the above changes have often brought with them a number of significant problems. These problems, which vary in character and degree from country to country and region to region, include concerns about:

-- the human health effects of pesticide and fertilizer residues, heavy metals, feed supplements and other contaminants in soil, water bodies, food products and the food chain;

-- the diminution and partition of biotopes valued for nature conservation;

-- the contamination of ground and surface waters and the eutrophication of surface waters by nitrates and phosphates leading to local health risks, declines in the quality of aquatic resources, losses in recreation values and increased water supply costs;

-- agricultural pollution problems associated with the growth of intensive animal husbandry;

---

* Pesticides are understood to include insecticides, herbicides and fungicides.

-- air pollution from intensive animal production, manure spreading and crop spraying;

-- the salinization of soils which is contaminating water supplies and causing losses in soil productivity and landscape amenity values;

-- losses in landscape amenity and wildlife habitat caused by the amalgamation of farms, the growing emergence of monocultures, the removal hedges, walls and terraces, the draining of wetlands and the deterioration and destruction of traditional farm buildings; and

-- soil compaction, erosion and pollution which have led to productivity losses, declines in the quality of water resources and reduction in the capacity of water storages.

At the same time the policies of other sectors and the pollution which has resulted from some of them have had adverse effects on agriculture in some regions. Severe financial problems have also arisen within the agricultural sector. For example, as a result of many of the price support policies designed to promote agricultural production, substantial agricultural surpluses exist, the resource intensity of agriculture is increased and the costs of agricultural support have escalated dramatically. Despite extensive state intervention, farmers in most OECD countries face serious financial difficulties and generally the outlook is not favourable.

The structural evolution associated with these changes has altered the rural landscape. The crisis in the agricultural sector and concern about the condition of our rural environment have been reflected in a number of international reports including:

-- the Commission of the European Communities has issued a "Green Paper" and follow-up guidelines for changes to the Common Agricultural Policy which include a number of initiatives to encourage structural adjustment (European Community, 1985a&b);

-- the World Commission on Environment and Development (1987) has found that agricultural subsidies and trade barriers in industrialised countries "have encouraged the overuse of soils and chemicals, the pollution of both water resources and foods with these chemicals and the degradation of the countryside";

-- OECD Ministers, noting that their economies are hampered by major distortions and rigidities in the agricultural sector, declared "that a gradual and balanced reduction in assistance to agriculture through a change in domestic support policies would be mutually beneficial to all Member countries and that this should be carried out on a multilateral multi-commodity basis"; and

-- Ministers at the Uruguay round of multilateral trade negotiations concerning the General Agreement on Tariffs and Trade (GATT) declared that there is an "urgent need to bring more discipline and predictability to world agricultural trade by correcting and preventing restrictions and distortions, including those related to structural surpluses, so as to reduce the uncertainty, imbalances and instability in world agricultural markets".

All these documents demonstrate that there should be a scaling down of support to agriculture. But as yet there is disagreement about how to achieve these reductions.

As a result of the above, significant structural change within the agricultural sector can be expected in the near future. Suggestions for the most desirable direction for this change vary from country to country and region to region. But in several countries restructuring is increasingly seen as compatible with a desire to improve the environment, promote regional diversification and reduce the dependence of certain rural areas on agriculture.

Opportunities beyond conventional agriculture which strengthen agriculture's contribution to the environment are being sought. Farmers are being encouraged to adopt, with appropriate incentives, a greater role in land care and management. Sources of farm income are being diversified to include off-farm income and investment. Part-time farming, for example, is being seen, in some countries, as an opportunity simultaneously to maintain landscape quality and facilitate the further development of tourism, recreation and decentralised industries and through it the revitalisation of the rural landscapes, of rural communities and of farming in general.

Positive contribution of agriculture to the environment

Agriculture's beneficial effects for the environment can only be considered in the light of a clear understanding of what is meant by a desirable environment. In the broadest sense, agriculture's greatest contribution to man's well-being is to provide the food and the "nutritional" security for the release of an increasing proportion of the population into productive activities in other sectors. At the same time, this has enabled man to enjoy increasing leisure time so that we can pause, enjoy and profit from our environment. Agriculture also has a special place in keeping the countryside inhabited, attractive and thriving. The presence of farmers and the maintenance of farm buildings can also improve environmental quality.

Over the centuries agriculture has shaped a landscape that is now highly valued and it is doubtful whether most people would wish it to be returned to its wilderness state. The demand for access to the countryside for enjoyment and recreation may be difficult to reconcile with a prosperous agriculture, but farming does keep open such opportunities. In most countries farmers are increasingly providing tourist and recreation facilities which increases the value of and returns from their land and their land maintenance efforts. More than half of Finland's farmers, for example, earn less than half their income from agriculture.

As a complement to conservation reserves and forests, it seems that most people's vision of an ideal countryside includes agriculture, albeit without certain of its recent manifestations. But there have always been conscious environmental trade-offs between agriculture's beneficial aspects and certain of its potentially negative impacts. Managed hedgerows, created by and for agriculture are visually more attractive and provide better wildlife habitat than wire fences. Their removal, to allow the introduction of labour-saving machinery, is not perceived by the general public as a benefit. Weedy, disease-infested, ill-tended crops are not desirable but weed killers and pesticides used to control these problems can have adverse effects on wildlife.

The drainage of land for farming has undoubtedly been a major cause of the elimination of malaria and probably other diseases in Europe and North America. Drainage and other soil cultivation operations can have diverse beneficial effects on flood control and water purity. Deep soils may moderate flash floods and paddy fields act as reservoirs and sediment settling basins. On the other hand, some forms of cultivation on hillsides encourage erosion, while other forms of cultivation can have possibly beneficial effects on the environment. Assuming that mankind will continue to consume animal products for the foreseeable future, farm land is essential to absorb the waste products of livestock raising. Man's own residues in the form of sewage sludge are also to a large extent absorbed by farm land.

Impact of pollution from other sources on agriculture*

If many of the mechanisms of environmental degradation associated with agriculture have only come to be more clearly understood in the past two decades, the state of knowledge about some of the major environmental influences on agriculture is even less advanced. Indeed, many of the phenomena now considered to pose significant long-term threats to agricultural production are only now coming under intense study. Among threats which have only relatively recently been perceived are:

-- increasing concentrations of CO_2 which, in addition to being a factor in the "greenhouse effect" of atmospheric heating, may also directly affect plant metabolism;

-- increases in ozone concentration at the earth's surface which appear to be causing declines in crop yield;

-- decreases in ozone concentration in the upper atmosphere, which allow greater amounts of ultraviolet radiation to reach the earth's surface, with as yet undetermined effects on plant growth;

-- pollution of the soil by harmful chemical and physical agents which has led to a decline in food quality and increased cancer risks; and

-- global climate changes, including prolonged periods of unseasonable weather and an apparent trend towards atmospheric warming which could have far-reaching effects on growing seasons and precipitation.

----

* This section summarises studies made by the ad hoc Group on Agriculture and Environment on the impact of pollution from other sources on agriculture. The work is summarised in greater detail in chapter 2 of the annex 1 to this report.

16

In other cases, the long-term cumulative impact on agriculture of well-known forms of pollution like water and air pollution from industry and sewage sludge spreading on agricultural land are only now being quantified. For example, European courts have recently recognized the damage caused to greenhouse owners downstream from potash mining operations which introduced large quantities of salts into the Rhine.

Although by definition unpredictable, nuclear and industrial accidents like Chernobyl and Seveso have occasioned major pollution of soil and water resources, loss of affected output and the need for significant financial compensation by governments.

In several countries, pollution from other sources has led to regional declines in the quality of food production and, in other cases, the quantity of food produced. Examples include damage to the quality of table grapes caused by acid rain in some parts of the Federal Republic of Germany and declines in milk production which in some parts of The Netherlands are believed to have been caused by fluoride emissions from secondary industry.

There is a growing body of literature describing the impact on agriculture of pollution from other sources. Most of it, however, is based on the extrapolation of the conclusions from laboratory experiments about the implications for regional and national crop estimates of acid precipitation, etc., on crop production. For example, a study in the United States estimated short-term crop yield losses due to ambient ozone concentrations at between 1.9 and 4.3 billion US dollars each year. Similarly, Japanese laboratory work suggested that, in some areas, acid rain and other forms of air pollution may be reducing wheat and rice crop production by as much as 30 per cent.

In summary, the nature, intensity, and extent of the acid deposition, photochemical oxidants and sewage sludge on agriculture remains largely unknown. Direct adverse effects on field-grown crops may be limited to a few sensitive species and in areas which experience severe local effects. The need for more research must be emphasised: cause-effect and, as a second step, consequential liming needs have still to be clarified at present levels of acid deposition.

2. GOALS AND EMERGING POLICY DIRECTIONS

Integration

Until recently, Member countries have tended to have separate sets of policies and usually separate departments to deal with environmental and other sector objectives. Today, however, it is becoming apparent that even though these objectives are sometimes in conflict, by recognising the impacts that each set of policies can have on other sector objectives, conscious trade-offs and hence more informed decisions can be made.

In particular, recent experiences are demonstrating that if more attention is paid to the long-term environmental consequences of implementing agricultural policies, substantial gains to the environment, to agriculture and other sectors are possible. Put succinctly, integration requires that full account is given to environmental objectives during the formulation of agricultural policies and, similarly, that environmental policies must reflect a recognition of their potential impacts on agricultural production, incomes and prices.

Integration necessitates the creation of institutional arrangements, the use of administrative procedures, and the formulation and implementation of policies which result in the more efficient and equitable achievement of related objectives. These objectives may be conflicting or complementary. Moreover, they can vary from region to region and be expected to change with time.

When these objectives are in conflict, integration is essential to ensure that the economically and socially "correct" trade-offs are made. Ideally, environmental and other sector policies should be mutually supportive and reinforcing in the long term. But in the real world this is not always possible. For example, while the continuation of traditional grazing practices is necessary for the conservation of certain species in France, in several agriculturally younger countries like Australia the introduction of cropping and grazing has been responsible for the decline and extinction of several native animals.

The development of integrated policies requires efforts to achieve greater complementary objectives and to make conscious trade-offs between competing objectives. It requires the development of policies that are preventative and anticipatory rather than reactive. Unintegrated policies are characterised by the belated recognition of their consequences for the objectives of other sectors.

There are three dimensions to integration: first, there is institutional integration which requires the development of administrative structures designed to ensure greater collaboration, co-operation and communication among agencies responsible for interdependent or related policies.

18

Second, there is the use of integrative procedures which include the development of agreed objectives; the soliciting of public comment on draft policy proposals and strategy plans; the use of environmental impact procedures; and the holding of public inquiries and the development of land-use plans.

Third, there is a whole set of integrative instruments, which through a series of interactions on each other can provide an optimal mix of trade-offs. Examples include the reallocation of property rights; the delivery of consistent advice from different departments to farmers on best management practices; the promotion of the concept of individual responsibility for the environment, the clear transmission of cost information in prices and the marginal cost pricing of water and other natural resources.

There are, however, costs associated with integration and the over-zealous pursuit of an ideal state of total integration may be counter-productive. Nevertheless, the types of benefits which can be expected from the effective integration of agricultural and environmental policies include:

-- economic gains to farmers in the form of greater opportunities for diversification and increased values of farm assets;

-- regional economic gains from a growth in regional tourism and recreation stimulated from improvements in landscape amenity;

-- social gains from improved public health, improved protection of wildlife, increased biological diversity and the maintenance of landscape amenity; and

-- national economic gains from more rational agriculture structures, the more efficient development and use of agricultural resources, and less expenditure on pollution control and nature conservation.

Agricultural and environmental objectives

Both agricultural and environmental policies and objectives are continuously changing in response to new circumstances. Although among OECD countries the broad policy aims for agriculture and environmental protection have tended to converge in recent years, considerable variation persists from country to country in the official sets of goals for agriculture and the environment. There is, however, a widely shared and growing recognition that within the OECD agriculture should give full consideration to, and should accept full responsibility for its effects on the environment. Similarly, those responsible for the environment should accept full responsibility for the effects of environmental policies on agriculture.

The words "agriculture and environment" have similar but subtly different meanings for each OECD country, due to the diversity of each country's cultural, physical and climatic conditions. In some parts of Europe, for example, care for the environment requires the maintenance of hedges, stone walls and traditional farm practices necessary for flora and fauna conservation. Other countries perceive care for the environment to include grazing to reduce fire hazard and terrace maintenance to reduce soil erosion.

In contrast, in Canada and Australia the need to preserve traditional farm landscapes is not perceived as an issue of major concern, whereas the protection of remnant native vegetation is.

At present, there seem to be four broad sets of environmental and agricultural objectives which can be observed in OECD countries. These are: a set of objectives associated with the maintenance of environmental quality; a set of agricultural objectives associated with producing food and fibre; a set of agricultural and environmental objectives which seek to achieve economic development and efficiency; and a set of income maintenance objectives.

Environmental quality

Human health

One important dimension of environmental quality is the maintenance of human health. Consequently, to ensure that the food products produced by agriculture are safe for human consumption, it is common in all countries to find restrictions on the use of pesticides, antibiotics, and growth hormones. Product standards such as the acceptable level of nitrates and heavy metals which can be found in food are also common. Generally, the principal health concerns in most countries are cancer and reproductive risks, applicator and farm worker safety, and the effects of chemical residues on the development of young children.

The maintenance of water quality is also a health issue. In recent years concern has risen about both the level of nitrates and nitrites in water and the noticeable pesticide residues which are being detected in some ground waters.

Nature conservation and landscape amenity

The value of the contribution which agriculture makes to nature conservation through habitat maintenance and plant cultivation depends upon the spatial distribution of these habitats and the extent to which the resultant biological networks interconnect. In agriculturally younger countries such as Australia, New Zealand, Canada and the United States few native species have adapted to agriculture, but in other OECD countries the reverse is the case and many species depend on specific agricultural practices for their survival.

Hedgerows and the grazing of mountain areas for example, are essential for the survival of several species, including many rodents and a large number of butterflies. Consequently, OECD countries are now implementing programmes to conserve hedgerows, conserve farm woodlands and retain the agricultural practices on which such plant and animal species depend. In contrast, the agriculturally younger OECD countries are introducing programmes to protect networks of native vegetation. In both cases, the objective is the retention of disappearing habitats and species valued for both ecological and economic reasons.

There is also a growing concern that the increasing tendency towards the development of highly specialised agricultural monocultures is causing significant losses in species abundance and diversity in many areas.

High quality rural landscapes are also vital for the development of tourism and for recreation and of course directly affect the quality of life for those who live in them. Consequently, in almost all countries, the loss of terraces, the loss of farm trees, the loss of traditional farm buildings, and creation of agricultural monocultures are considered to result in a decrease in landscape amenity and heritage values.

In several countries, maintaining the character of the rural environment is perceived to require the retention of farmers and the family farming system. In countries such as The Netherlands and Switzerland, tourist development policies are being used to exploit characteristic rural environments to advantage. Dutch tulips and dairy cows are now national symbols. Conversely, extensive low or sparsely populated agricultural areas are also widely appreciated for their open "wilderness" character.

Quality of natural resources

Farmers are dependent on the quality of their natural resource base. Soil conservation, pasture and forestry management, erosion control and maintenance of water quality are all an integral part of the wise use and conservation of natural resources. By adopting practices that contribute to rational use and conservation of natural resources, farmers can make a positive contribution to the environment in addition to contributing to agricultural objectives. Measures which give incentives to misuse these resources, by disguising their true value and cost, work against environmental quality.

Demands for the conservation of natural resources also exist because at a national level demands for high economic efficiency require economy in the use of agricultural resources such as soil and the maintenance of an appropriate ecological balance. Similarly, demands for economic efficiency require that pollution from all sources including agriculture be minimised. Such approaches keep future options open and tend to enhance the value of related national resources. Fishing, for example, is dependent on water quality and this, in turn, is improved by soil conservation.

Economic development and efficiency

The promotion of economic development and efficiency is an objective common to all sectors. Its roots lie with the concept of the benefits which can be expected if all countries, all regions and all sectors maximise the opportunities their comparative advantages give them. As pointed out in other OECD fora these benefits will be greatest when there is free trade and open but fair competition among nations. Amongst other things, this also requires the internalisation of external costs such as those associated with environmental pollution arising from agriculture.

The maintenance of economic efficiency requires countries to continuously exploit their economic, environmental, ecological and physical advantages and, as circumstances change, to adapt their agricultural production systems so that they remain efficient. Economic efficiency would ensure, in the long run, low food prices and, providing the costs of pollution are internalised, less agricultural pollution. The structural adjustment which would be accompanied by a move towards greater economic efficiency can be expected to increase the standard of living of those farmers who remain in agriculture, but also require a significant number to leave the industry. In some cases this may include the development of part-time farming and leisure farming. However, the overall gains in employment in other sectors of the economy resulting from the movement of the labour and capital resources from their less efficient uses in agriculture to their more efficient use elsewhere in the economy could offset agricultural employment declines.

If the environmental costs of agricultural pollution were internalised, this would lead to a different, more efficient distribution of agricultural resources. On the other hand, it must be recognised that the physical distribution of production is dominated, not only by comparative advantage, but also by a whole system of agricultural policies designed to alter production and trade policies in directions seen as nationally desirable.

Given the complexity of the problem of attaining economic efficiency and the paucity of progress towards a solution, it would be idle to pretend that governments will suddenly shift their efforts from intervening in the market system to protecting the environment. Without such a change, however, any economic changes in favour of the environment are likely to be swamped by trade-oriented policies.

While sorting out the trade problem would not automatically solve the environmental problem, there nevertheless are possibilities of convergence between agricultural trade improvement and environmental protection. In general, measures that create incentives to increase intensity in agricultural production also lead to increased stress on the natural environment.

Along with this more formal economic advice, there is a growing public perception that it is not logical to damage the environment to produce surpluses. Some damage may have to be accepted to avoid food shortages but when the marginal value of food is zero or even negative it is irrational.

Income maintenance

Almost all OECD countries give high priority to the maintenance of agricultural incomes. Views range from the conviction that "farming is the backbone of the nation" to a vague feeling than a fall in agricultural incomes would lead to a flood of "peasants" onto already over-charged labour markets. More recently the perception has emerged that keeping farmers on the land, especially in less favoured regions, is a desirable national goal in its own right. Some believe that farmers occupy a special position in society and are seen in many countries as in need of income support and other financial assistance. Other sectors rarely enjoy the same consideration. The reasons for this objective are complex, but it is a dominant one and lies at the root of policies which in some countries are designed to maintain stable agricultural prices at levels which are well above world market prices.

Whatever the merits or otherwise of direct and indirect income support policies, they clearly have not succeeded in maintaining farm income, nor farm populations at pre-support levels. A discussion of the employment effects of agricultural support is outside the scope of this study. Nevertheless, even if support policies have been shown to be rather unsuccessful in maintaining farm incomes, there can be no doubt that the sudden introduction of environment protection measures can have negative effects on farm income. Conversely, it is recognised that ad hoc or haphazard forms of radical change in agricultural policy could have an adverse effect on the environment.

3. APPROACHES TO INTEGRATION

The focus of the remaining sections of this report is on identifying successful approaches to and opportunities for the integration of agricultural and environmental policies. The methodology used in identifying these approaches and opportunities was to collect information about and from countries, commission a series of detailed policy area studies and, where necessary, supplement this with information from other sources. This approach has resulted in two reports: one on "Country approaches to the integration of agricultural and environmental policies", and another on each of the specific policy area studies.

In light of the above two reports the first part of this chapter presents only a brief overview of the first report on country approaches, and the second part deals with the major policy implications which arose from the policy area studies. Appropriate additional information from other sources is also included. Section 4 then discusses some major concepts which are relevant to improving integration so that sections 5 and 6 can highlight existing and emerging opportunities for improving integration.

International, national and regional approaches to integration

Figure 2 outlines the nature of the different approaches to the integration of agricultural and environmental policies at the institutional, procedural and policy instrument level.

Institutional integration

Institutional arrangements for the integration of agricultural and environmental policies vary tremendously among OECD countries. The reasons for this are complex and depend for a large part on cultural and historical factors. A marked feature of the approaches, however, is the extent to which new innovative structures and arrangements have been developed within a relatively short time frame.

At an international level, bodies such as the Food and Agricultural Organisation (FAO), provide a forum for international debates, establish general principles and set standards. One example of this is the FAO's Codex Alimentarius Commission which sets maximum pesticide residue limits and quality standards for food in international trade. FAO also plays a leading role in helping countries to regulate and use pesticides. Others, such as the European Community via its directives and general co-ordinating role for its twelve Member countries, has a dominating effect on the quality of the rural environment and location of agricultural production.

At a national level, the historical place of agriculture has meant that ministries and departments of agriculture have existed for decades in all OECD countries. Departments of the environment tend to be more recent, however, and arrangements for the co-ordination of environmental and agricultural activities less well established.

Figure 2

COUNTRY APPROACHES TO INTEGRATION

ADMINISTRATIVE INTEGRATION

-- National
  * Create Environment Ministry
  * Transfer Responsibility from the Ministry of Agriculture to the
    Ministry of Environment
  * Assign Joint Responsibilities

-- National and regional level
  * Create Laws which require integration

INTEGRATION PROCEDURES

  * Change policy formulation process
  * Public participation
  * Inquiries, task forces and working groups
  * Land-use planning
  * Environmental impact assessment

INTEGRATION INSTRUMENTS

-- Advisory approaches
  * Direct advice to farmers
  * Education via media, etc. to farmers and general public
  * Farmer initiated conservation schemes

-- Economic approaches
  * Input taxes
  * Implementation of the Polluter-Pays Principle
  * Land set-aside
  * Direct conservation payments
  * Removal of subsidies

-- Regulatory approaches
  * Chemical standards (Fertilizers & Pesticides)
  * Restrictions on potentially polluting agricultural practices
  * Prohibition of undesirable agricultural practices
  * Licensing requirements

Over the past two decades, most OECD countries have created agencies or ministries specifically charged with the development and implementation of policies and programmes to monitor and protect the environment. These developments suggest that changes in administrative arrangements and administrative procedures are often a necessary precondition to the successful integration of agricultural and environmental policies with those of other sectors.

Co-ordination among ministries, departments, national and regional, and local governments is achieved through a broad cross-section of mechanisms. A noticeable recent development has been the growth in mandatory inter-agency consultation procedures and formal inter-agency agreements for collaboration in the development and implementation of policy.

In many countries, responsibility for the rural environment still remains almost exclusively with the agricultural authorities. Nevertheless, once environment ministries have been established, the trend has been towards the progressive transfer of responsibility from agricultural to environmental agencies.

When several departments are responsible for the management of the rural environment, it is difficult to clearly establish the limits of each department's role. Recognition of this problem has led some countries to redefine and reassign responsibility for certain joint agricultural and environmental objectives to a specific agency. This approach of creating agencies with specific responsibilities and very clear objectives, is most effective when the agencies remain accountable to the public and are highly visible.

Another approach has been the establishment of clearly identified environmental units within agricultural ministries. These units usually have responsibility for ensuring that all the likely effects of agricultural policy on the environment are considered. Units with responsibility for rural affairs are also beginning to emerge in environment departments and, as a result, transfers and exchanges of expert personnel between agricultural and environmental agencies are becoming more common.

Administrative arrangements are also being changed to ensure that environmental questions are given full consideration in agricultural research. In many countries, it is now common practice to appoint people with environmental interests to boards and committees allocating funds for research. Monitoring arrangements are being changed also so that the data collected can make a greater contribution to agricultural policy formulation.

Integration procedures

The need to integrate agricultural and environmental policies appears to have become formally recognised by governments in the 1970s. Today, policy statements on the need to ensure that farmers and agricultural policy continue to work to maintain and enhance the quality of the environment can be found in most OECD countries.

Recognition of the interdependence of agricultural and environmental policies has led to the revision of procedures for policy formulation. Public inquiries are becoming more common as is the involvement in decision-making of the plethora of groups interested in agriculture and the environment including farmers' associations, consumer groups, environmental groups, farm suppliers, the media and the scientific community. Non-profit foundations, special interest groups and fully-fledged ecologically orientated political parties all now have some capacity to influence policy in most OECD countries. In some countries these non-government organisations are now routinely and actively encouraged to participate in policy formulation and development.

Committees, task forces, working groups and issue specific inter-agency groups with responsibility for finding integrated solutions to agricultural and environmental problems can now be found in virtually all OECD countries. In recent years a number of countries have moved to develop a series of programmes which fully integrate agricultural and environmental policies. Characteristically, these programmes tend to address the causes, sources and effects of environmental problems while recognising the need to attain agricultural objectives and acknowledging that other industry activities also affect the quality of the environment.

One excellent example of this is the Dutch working group of three ministries which produces a revolving multi-year indicative programme for agriculture and the environment. This group has also developed a two-track "source-effect oriented" approach. Agriculture is one of the target groups for source oriented policy and measures are being taken to discourage practices which cause problems such as nitrate pollution. Effect-oriented policies, on the other hand, are being used to reduce existing environmental problems.

Experience in several Scandinavian countries suggests that in resolving agricultural pollution problems the careful negotiation of clear, concise, agreed policy targets and objectives, beginning with the definition of the trade-offs to be made between agricultural and environmental objectives is a more successful approach than the more traditional one of negotiating a set of policies to address an environmental problem without clearly defining the preferred solution as an objective. Such identification of targets and objectives avoids preoccupation with problems associated with overlaps at boundaries of administrative responsibility which must be expected to remain no matter what policies are adopted.

A different but equally successful approach, which is growing in acceptance throughout the OECD, is the development of water catchment plans which guide all forms of land use within a catchment area. Regional land-use planning, particularly when backed by thorough natural resource studies, inventories and accounts, can also be extremely useful in facilitating the development and implementation of integrated policies.

Another administrative procedure which can be used to further the integration of environmental and agricultural policies is environmental impact assessment. In the past, environmental impact assessment has been most often applied to the planning of large-scale industrial and infrastructure projects in OECD Member countries. The scope of environmental impact assessment is being continually broadened, however, and through initiatives such as that taken by the Commission of the European Communities, agricultural projects may be subjected to environmental impact assessment in the future. The Communities' list of agricultural projects which could be assessed for environmental impact include: the restructuring of rural land holdings; the use of land for intensive agricultural and animal husbandry purposes; and afforestation where this may lead to adverse ecological changes.

Legislation can also play a major role in promoting integration especially when it is used to specify the nature of agreed objectives and outline mandatory administrative procedures. There is a risk, however, that if legislation is too prescriptive, it can hinder development and the adoption of more efficient and more effective administrative procedures.

Integration instruments

Advisory and voluntary approaches

OECD Member countries are almost unanimous in a preference for advisory and voluntary approaches to integration backed, as a last resort, by appropriate regulations. The adoption of environmentally appropriate practices by farmers relies on encouragement and persuasion. This approach also involves the development of public education, which by changing peer group pressure, is proving to be extremely effective in strengthening community attitudes to the environment. "Land care" or "land stewardship" is promoted as a moral responsibility, but it can not change the fundamental economic price and cost structures which determine whether or not farmers can profitably adopt the advice given.

Education programmes at both the tertiary and secondary levels are also being consciously changed to make advisory officers, the public in general and farmers more aware of the benefits of adopting environmentally sound agricultural practices.

In the case of non-point sources of pollution such as soil erosion, it is very difficult to identify the farmers who are causing the pollution problem. Hence, in these areas the emphasis is usually very strongly directed towards the provision of advisory and extension services which promote environmentally sound agricultural practices. This advisory approach can bring long-run economic gains for the farmers, as well as the environment, from a reduction in the over-use of inputs and improved soil productivity.

There can be problems in ensuring that different advisory activities are integrated so that farmers are not receiving contradictory messages. This has led in recent years to a rationalisation of extension services so that the greatest effort is targeted to priority problem areas. It has been found in some cases that by targeting extension and advisory resources towards areas where, for example, soil losses and damages are greatest, extension becomes a very effective method of controlling non-point sources of pollution. Targeting also tends to increase the environmental skills of the advisory officers involved.

In some countries, such as Australia and Canada, farmers are starting to develop their own conservation schemes. This is especially apparent in areas where at the local and regional level government participation in farmer conservation schemes is seen as being much more successful than farmer participation in government schemes. One way of achieving this is to appoint, or ask farmers to elect, a leading farmer as chairman of the implementing body, which may even contain a majority of farmer members. Government staff are increasingly seen as specialist collaborators rather than authoritarian "imposers" of land protection schemes. Their role is becoming more and more one of motivating individuals and communities and, once interest is aroused, specialist advisors.

Advisory approaches are heavily dependent for information on the conduct of effective research. The agricultural and environmental research being conducted in countries is often not designed in a manner which seeks to address the environmental problems associated with agricultural production and, conversely, the effects of the environmental pollution on agricultural production. Increasingly, however, countries are promoting research into alternatively forms of agriculture which use less inputs which have negative effects on the environment.

Economic approaches

Farmers tend to be responsive to changes in the relative profitability of different products and the cost of practices used to produce them. Consequently economic instruments are often used to influence agricultural production for a variety of objectives such as a desire to become self-sufficient and a desire to increase national productivity and exports. When such policies act by supporting prices, they cause an increase in the intensity of production and can thus lead to overuse of natural resources and chemical inputs and to increased run-off and groundwater pollution. Economic instruments are also used to maintain and improve the environment. Cost sharing, for example has been used to encourage farmers to conserve wetlands and other natural resources. Another approach is to acquire an area of land and then lease it back to the farmers under suitable restrictions on use. A third is to acquire a conservation easement over an area of land.

One barrier to successful integration can be the level of financial assistance and protection for agriculture is substantial in most OECD countries. As a result the intensity, location and type of practices found in almost all agricultural regions are not arbitrated by market forces. The level of assistance varies widely across commodities and is provided through a large number of policy instruments. Measures used to increase the prices received by farmers include the provision of deficiency payments, direct market intervention and the imposition of tariffs and import quotas. Other economic approaches include the provision of grants and concessional loans; the use of input subsidies and taxes; and levies on product sales. All of these measures distort the cost information transmitted through prices. Such policies promote the intensive use of resources while causing undervaluation of the costs of production which include the off-site cost of preserving common property resources such as water.

Some of these instruments have positive environmental effects, while others have negative effects. Moreover, the nature of these impacts changes from region to region and sector to sector. Almost without exception the negative effects have been unintentional. With hindsight, many governments are now seeking ways to enhance the role of such instruments in producing environmental benefits while at the same time achieving traditional agricultural objectives.

Product price and other arrangements designed to increase production

Although rarely used as an instrument for achieving integration, changes in product price and other arrangements designed to increase agricultural production have a significant influence on environmental quality. Two different approaches to product price arrangements can be observed amongst OECD countries. The first approach is to remove all price support arrangements and expose all sectors of the economy to market forces. The expectation is that this will create a more efficient agricultural sector and in turn lead to a reduction in pressure on the environment as agriculture becomes a) more extensive in the sense that the use of environmentally negative inputs per hectare will become lower; and b) more diversified in the sense that farmers will grow a greater variety of products. Advocates of this approach argue that in the long run, higher prices increase production intensity and, through the greater use of inputs, a correspondingly higher risk of pollution. The second approach begins with the notion that product price and other assistance measures are necessary to ensure that farmers receive adequate incomes, that regional economies are developed, that self-sufficiency is achieved and that landscapes in less favoured regions are maintained.

Some price support schemes have distorted the capital structure of agriculture and created unwanted production surpluses. In recognition of this problem, many countries are now seeking ways to reduce the intensification of agriculture and take land out of production. In certain circumstances, policies which reduce production, if properly targeted, can have substantial benefits for the environment. Such policies include price reduction and the use of input taxes and charges to finance the sale of surplus grain and price schemes which offer a lower price for over quota production of milk. Transferable quotas allow structural change by allowing agricultural production to shift to regions where it has less environmental impact. The merits of allowing such a shift need to be considered against its impact on regional economies and the environment. Consideration also needs to be given to the land use that is likely to occur as a result of quota transfers.

Several countries are concerned about the effects on the environment of land reverting from extensive crop production to grazing land or, alternatively, being withdrawn from production. Others are implementing a variety of schemes which pay people to take land out of production. Typically, the newer versions of these schemes have both welfare or income support and environmental objectives.

Land set-aside programmes of various types have been discussed and tried in the past with varying degrees of success and within the European Community it has now been decided to introduce set-aside as one way of helping to reduce current agricultural problems. If farmers take crop land out of production, for say 10 years, there is reason to expect habitat change with benefits to flora and fauna on such lands, but if the land is only set aside for one or 2 years the experience has been that the set-aside leads to further environmental deterioration. This deterioration is particularly severe when the land is left as bare fallow and exposed to the risk of increased soil erosion and the increased leaching of nutrients.

As well as creating surpluses in some situations, product price support policies are used to change land use patterns. Higher prices may be paid to maintain regional production and employment. This policy has also prevented parts of the landscape from reverting from agriculture to forestry. In other marginal areas there have been calls for similar differential price arrangements to maintain landscapes in less-favoured areas.

An alternative policy option which is being widely discussed within the European Community and also in other fora is to reduce the level of price support and provide income support payments to farmers in a manner which does not encourage production.

Input subsidies, charges and levies

The principal economic methods of deriving environmental benefits from agriculture have traditionally involved subsidies, concessional loans and grants which encourage farmers to adopt environmentally favourable practices.

Input taxes and charges can produce environmental benefits by decreasing fertilizer and pesticide use. This decrease not only results from the higher cost of these inputs but also because it makes farmers review their whole fertilizer policy and, for example, stimulate greater care in the use of animal manure as a fertilizer. They change farmers' perceptions of fertilizer costs and yield responsiveness, leading to a recognition that profit maximisation often requires the judicious use of inputs.

To protect native flora and fauna in countries like Germany, farmers are paid to leave pre-determined crop edges unsprayed when they apply pesticides and also to leave the edges of fields uncropped. Payments are also offered to farmers who refrain from spraying pesticides altogether, reduce levels of fertilizer or leave meadows unused during main insect hatching periods.

In most countries water for irrigation is generally either subsidised or free. Low water charges tend to cause land to be used more intensively and more water to be used per hectare than would be the case if it were priced at its true economic cost. Cheaply priced water often reduces incentives to adopt technologies such as land forming which, through the more efficient use of water, would reduce salinity and downstream water management costs.

Tariff policies and preferential trading agreements may also affect the use of inputs and intensity of agricultural production.

Generally the experience with varying the price of agricultural inputs so that they take account of the environmental costs of agriculture has been favourable, although in most cases and unless supported by other measures very high input taxes are needed to significantly reduce pollution. Increasing the cost of agricultural inputs makes some farmers more aware of the additional profits which can be made by the more efficient use of inputs and also, reduces both on and off-farm pollution by internalising some of the off-site costs of pollution from agriculture. Where input prices do not reflect their true on and off-site costs, appropriate regulations and standards may be needed to prevent environmental problems.

Regulatory approaches

Generally OECD farmers have freedom of choice in the selection and use of inputs, subject to restrictions only where these are necessary to meet product, producer and environmental standards. Standards are normally set for all forms of agricultural chemicals. For example, pesticides must be licensed and standards are set for the composition of fertilizers. The use of hormones and specialised additives to animal feed stuffs is also strictly controlled in each country. Procedures for the examination, licensing and review of these chemicals, however, vary considerably among countries and chemicals banned in some countries are still used in others. Extensive but differing labelling standards also exist in most countries. Regulations on the application of chemicals, however, are often either non-existent or not enforced.

Following the development of analytical methods which make it easier to detect trace elements, people have begun to express concern over the heavy metals accumulation in soils and pesticide residues in water. Some, but not all OECD countries, have set limits on the maximum amount of cadmium which may remain in mineral fertilizers sold to farmers.

Regulations are also used in OECD countries to control agricultural development so that it is compatible with environmental objectives and to ensure that as far as possible agricultural activities make a positive contribution to the environment. There is, however, a noticeable reluctance on the part of many countries to enforce land-use restrictions to maintain environmental quality. Zoning is used in most countries to control agricultural development by defining permitted new uses. Regulations are also used to protect agricultural land from urban and industrial encroachment.

For predominately socio-political reasons, some countries also regulate land ownership in favour of family farms. Most of these countries consider that "family farms" are more environmentally favourable than other types of farms. One method of doing this is to place restrictions on the transfer and amalgamation of farms.

There are few controls over farm practices in most OECD countries and, in many cases where controls exist, they are very difficult to enforce in a cost effective manner. The exceptions are nearly all associated with animal cruelty and point sources of pollution which have significant off-site costs. One such practice is the burning of straw. Another is the spreading of manure and sewage sludge. Several countries are also requiring farmers to build animal manure storage facilities capable of storing manure for 6 to 9 months.

Most countries with nitrate pollution problems have or are providing subsidies to assist farmers to upgrade these facilities during the transition period when more stringent regulations are introduced. Similarly, in most countries, it has become common to zone sensitive water catchment areas and to restrict land-use practices within them. Outside water catchment areas, it is extremely rare for farmers, however, to be prosecuted for not complying with land-use regulations. In many cases, appropriate penalties and enforcement activities are non-existent.

Product marketing regulations, through their effects on agricultural practice, can also effect environmental quality. Product grading standards, for example, influence the use of pesticides and as a result changes in grading standards in the United States are now subject to environmental assessment when they are thought to have a significant effect on the environment. Similarly, if the recent growth in organic farming continues, the regulations and standards for the production, marketing and labelling of organic products can be expected to have an effect on environment quality in certain regions. Countries like Denmark have recently introduced organic food production regulations and the Commission of the European Communities is in the process of developing a uniform set of regulations for its twelve members.

Lessons from specific policy areas for integration

Intensive crop production and the use of agricultural chemicals*

This policy area focused on the environmental implications of the expansion of intensive crop production and the increasing use of agricultural chemicals in most OECD countries (Figure 3). Amongst other things, these increases have been due to a combination of plant breeding advances; technological developments which have facilitated the use of heavy and sophisticated machinery; changes in labour and other factor costs; and government price support for major agricultural commodities. Taken together, these factors have caused substantial structural change. The result has been that crop production has become increasingly specialised and more intensive in terms of agricultural chemical and equipment use and more widespread in areas not always suitable for cropping.

The nature of the environmental impacts which have resulted from this structural change have been similar across most OECD countries. The incidence of these impacts, however, varies substantially from country to country. The most noticeable are the reduction of species caused by the diminution and partition of natural biotopes and habitats; the contamination of groundwater by nitrate and pesticides; the compaction and erosion of soil; the eutrophication of surface waters; the increased detection of pesticide residues in food; and air pollution from dust and odours produced by certain agricultural practices. Many of these impacts are due to the increased use of agricultural chemicals. Nitrogenous fertilizer use has more than doubled in

---

*   The policy area studies on "Intensive Crop Production and the Use of Agricultural Chemicals" in Germany, in Sweden and in the United States, are summarised in greater detail in Chapter 2 to the annex 2 to this report.

Figure 3

NITROGENOUS FERTILIZER USE IN OECD COUNTRIES

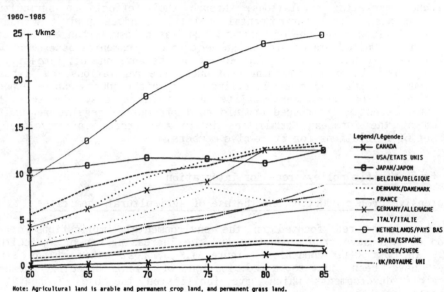

APPLICATION OF NITROGENOUS FERTILIZERS, In tonnes per km2 of agricultural land

1960-1985

Legend/Légende:
- CANADA
- USA/ETATS UNIS
- JAPAN/JAPON
- BELGIUM/BELGIQUE
- DENMARK/DANEMARK
- FRANCE
- GERMANY/ALLEMAGNE
- ITALY/ITALIE
- NETHERLANDS/PAYS BAS
- SPAIN/ESPAGNE
- SWEDEN/SUEDE
- UK/ROYAUME UNI

Note: Agricultural land is arable and permanent crop land, and permanent grass land.

many OECD countries in the last 25 years. For example, in Denmark fertilizer use has increased by 225 per cent, in the United States it has increased by 300 per cent, and in The Netherlands it has increased by 150 per cent. The frequency and quantity of pesticide use has also increased in many countries. For example, since 1975 the quantity of pesticides used has increased by 69 per cent in Denmark and, 30 per cent in Germany. Similarly, the frequency of application of pesticides increased 115 per cent in the 3 years between 1981 and 1984 in Denmark (Table 1).

Table 1

PESTICIDE USE IN OECD COUNTRIES

CONSUMPTION OF PESTICIDES, OECD countries, early 1980s

Tonnes (active ingredients)

	Year	Pesticides total	Insecticides	Fungicides	Herbicides	Other pesticides
Canada	1985	39259	3172	2823	30181	3083
USA 1)	1981	334000	41000	25000	268000	-
Japan	1985	83056	45018	18622	19416	-
Australia	-	-	-	-	-	-
N. Zealand	-	-	-	-	-	-
Austria	1984	4845	439	1739	2428	240
Belgium	-	-	-	-	-	-
Denmark	1984	8018	437	2407	4702	473
Finland2)	1985	1981	165	110	1641	65
France	1982	93400	5500	56700	31200	-
Germany3)	1985	30053	1566	8491	17390	2606
Greece 4)	1976	29940	2695	26348	897	-
Iceland	1981	6	1	3	2	-
Ireland	1980	1470	185	210	1050	25
Italy 5)	1983	155946	33188	82004	26056	14699
Luxembourg	-	-	-	-	-	-
Netherlands	1985	19938	634	4363	3977	10964
Norway	1982	1527	36	92	1332	68
Portugal	1982	14007	440	12506	955	106
Spain	-	-	-	-	-	-
Sweden	1984	4946	186	859	3760	141
Switzerland	1982	2046	131	1085	830	-
Turkey	1982	8119	4619	1575	1925	-
UK 6)	1982	40300	1480	4780	28100	5900
Yugoslavia4,5)	1976	25384	10448	9477	5459	-

NOTES:

1) Secretariat estimates based on partial data.
2) Includes quantities used for forestry.
3) Provisional data.
4) Data refer to the late 1970s.
5) Data refer to the formulation weight.
6) Great Britain only. Data refer to the early 1980s.

Source: FAO, national statistical yearbooks.

Examples of the nature of the impact which this increased use is having on the environment include:

-- the emergence of significant areas where the groundwater exceeds the European Community guideline of 25 mg/l. of nitrate per litre of potable water and also the maximum limit for potable water of 50 mg/l in countries like Denmark, the United Kingdom, Sweden and Germany;

-- the loss in the one part of Germany of 36 per cent of hedgerows between 1954 and 1971, and 50 per cent of the remainder were lost in just 8 years between 1971 and 1979 (Figure 4);

-- the potential contamination of groundwater suggested by the recent detection of 56 different pesticides in some groundwaters in the United States;

-- the severe eutrophication and oxygenation of coastal waters in Scandinavia; and

-- in the 33 fold increase in the rate of species extinction in the intensive crop production area of Niedersachen in Germany where only 14 species became extinct in the 80 years between 1870 and 1950, but 131 species became extinct between 1950 and 1970 (Table 2) with eighty-five per cent of these post-1950 losses being attributed to agricultural practices.

Table 2

NUMBER OF DISAPPEARING PLANT SPECIES IN
THE INTENSIVE CROP PRODUCTION AREA OF NIEDERSACHSEN IN GERMANY

Period	Number of species
1870 - 1950	14
1950 - 1978	131

Source: Annex 2, Chapter 2

Based upon the above and many other examples, the work in this policy area found that the existing range of agricultural policies, coupled with recent technological developments has led to inefficient and environmentally damaging levels of agricultural chemicals use and that action to reduce the extent of use and adverse effects of these chemicals is necessary.

In several countries there is mounting evidence of the tendency of farmers to use agricultural chemicals at above the level which is optimal for profit maximisation. In Sweden, for example, one survey found that by applying the fertilizer necessary only for profit maximisation and making maximum use of animal manure, net farm income could be increased by 100-150 Skr per hectare (Figure 4). Similarly, a recent United States survey found that over half the nitrogenous fertilizers applied in the corn belt was not needed to achieve maximum profits. Some reasons for this observed tendency of farmers to over-utilise fertilizers and pesticides in these countries are that:

-- the costs to the farmer of marginal increases in application rates are small while fixed costs, particularly application costs, form the major proportion of total costs;

-- response rates for individual fields are not known precisely and vary with external factors, such as weather conditions and disease attacks making the over-application of chemicals the optimal risk averse strategy;

-- farmers find it too expensive to utilise animal manure or are unaware of its fertilizer value;

-- the livestock density on individual farms is too high;

-- peer pressure, through judgements of "good husbandry", create social pressures for farmers to produce weed and disease-free, high yielding crops;

-- production incentives and guaranteed market prices which reduce the risks of high input use in many countries;

-- marketing regulations, government assistance programmes, product guarantees, etc. sometimes require the use of certain practices which have adverse effects on environmental quality; and

-- food processor requirements that farmers use certain pesticides during the production of fruit and vegetables.

Figure 4

EXISTING AND RECOMMENDED FERTILIZER RATES FROM A SURVEY
OF 30 FARMS IN SWEDEN

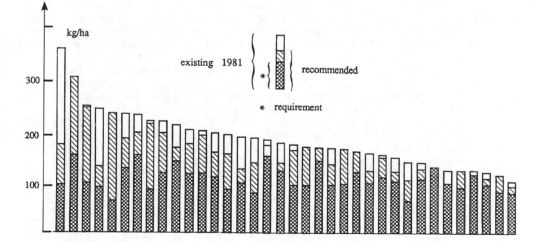

Policy options being considered and evaluated by Governments in an attempt to reduce the magnitude of the above problems and impacts include:

-- strictly enforcing and strengthening existing environmental regulations;

-- imposing new stricter environmental regulations;

-- controlling production through decoupling income support from price support;

-- controlling production through eliminating target prices and reducing the total area cultivated;

-- establishing reserves from diverted cropland;

-- reviewing current research programmes and developing new ones which seek to reduce pollution associated with the use of agricultural chemicals; and

-- reducing the intensity of cultivation practices via input taxes.

In Sweden, for example, input regulations are also being strengthened. To register new pesticides, companies are now being required to test and provide information on the consequences of applying pesticides at a level which gives farmers less than full guaranteed protection. Extension officers are being encouraged to utilise this information and to promote the adoption of integrated pest management techniques. They are also being encouraged to promote the use of agricultural practices which reduce nutrient leaching, such as the planting of autumn catch crops to take up nitrogen after harvesting, and encouraging manure spreading to the spring when there is maximum plant growth.

The study also found that in several countries there are few land use regulations and those which exist are poorly defined and unenforced. One way of overcoming this problem is to clearly define what is meant by "good farm practice". In the United States best management practices are being defined to overcome this problem at a local level, but as yet, the use of these definitions as a means of facilitating land-use control is rare. This situation is changing, however, and the Commission of the European Communities, for example, is seriously considering introducing restrictions on the maximum amount of nitrate from all sources which may be applied per hectare on a crop by crop basis. Similarly, in Denmark, new land-use regulations have been introduced to reduce nitrate and phosphate pollution. Under these new regulations the planting of autumn catch crops is required by law and farmers must prepare fertilizer management plans which indicate how they will apply fertilizers and spread manure in a manner which does not cause unacceptable water pollution. Under these new catch crop regulation 45 per cent of the area of each farm must be under a green cover by autumn 1988. This regulation, which seeks to reduce nitrate leeching from bare fallow lands, will be increased to 55 per cent in 1989 and 65 per cent in 1990.

In several countries it has now become clear that the simple enforcement of current, and even considerably strengthened regulations, coupled with the introduction of stronger advisory services is unlikely to sufficiently reduce pollution from agricultural chemicals. As a result, the use of economic instruments and quotas are being evaluated. These instruments can operate on both input and output prices and, within this context, policies which seek to integrate environmental and agricultural policies are being considered.

Two types of input taxes and levies are being discussed by countries. Relatively low input taxes and levies in the vicinity of 25 per cent are being used in countries like Sweden as a means of partially financing extension and advisory activities and, also, research into the reduction of agricultural pollution. Farmer acceptance of these levies is reported to be high as far as the funds raised are used to support export costs. Experience with similar input taxes also exists in Finland and Austria. At low rates, however, their direct effect on consumption appears to be negligible.

No country has yet introduced high input taxes, but, in several, taxes of the order of 100 per cent to 200 per cent have been discussed. Such taxes would have a substantial impact on farm income and are also vulnerable to changes in input and output prices. In Germany, for example, it was estimated that a 200 per cent fertilizer tax would reduce the use of mineral fertilizers by 30 per cent and farm income by 25 per cent. At the same time, however, a 200 per cent tax could only be expected to reduce water pollution by 50 per cent as farmers would adjust their crop rotation patterns to include nitrogen fixing legumes. In contrast with fertilizers, however, a 200 per cent tax on pesticides was found to reduce consumption to a mere 18 per cent of its previous level, with an increase in labour requirements as mechanical weed control replaced chemical control. A related problem is that high taxes inequitably penalise farmers who use fertilizers in a manner which does not contribute to pollution.

Many of the adverse effects on farm income, together with inequity problems could be overcome by introducing transferable input quotas at a regional level. Although input quotas have been discussed and even proposed by farmers' organisations on several occasions, no country has attempted to use input quotas as a means of reducing agricultural pollution. Nevertheless, the German study found that input quotas have less effect on farm income than input taxes and, if they are restrictive, provide a strong economic incentive for farmers to use inputs efficiently. They are also not sensitive to price fluctuations and are easier to enforce than land-use regulations. Quotas on the total amount of inputs which a farmer may use have the advantage of addressing the environment problem directly. They can be made region-specific and farmers can be given a right to a certain level of chemical use, associated with a reasonable crop loss risk level.

An alternative and complementary approach is to vary production incentives in a manner which helps to overcome both the pollution control problems associated with agricultural chemicals and the agricultural surplus problems which exist in many countries. The German study in this area found that the environmental effects of reduced production incentives are similar to an input tax. For the same reduction in farm income, however, the reduction in pollution from agricultural chemicals would be considerably less than that which would be achievable via input quotas and taxes.

One option is to reduce price support and provide direct income support to farmers. In many OECD countries cereal prices are well above world prices and production significantly exceeds consumption, necessitating high export subsidies to dispose of surplus grain. The result is a wide gap between the farm-firm cost of any production reducing measure and the socio-economic cost. This situation has led to discussions about the merits of decoupling price and income support policies. From the viewpoint of pollution from agricultural chemicals the substitution of direct income support policies for price support would, in many areas, by reducing economic incentives for the application of agricultural chemicals reduce pollution and be environmentally advantageous in areas where pollution is a problem.

Another approach is to pay farmers to set aside land or use less inputs as a means to reduce agricultural surpluses and improve the environment. If properly targeted and administered, so as to produce environmental benefits, such policies could be financed from budget savings. Work in the United States which has examined the impact of targeting such schemes has found that the budgetary savings from such targeted schemes can be substantial. For example, it has been calculated that a 20 million acre expansion of the existing conservation reserve programme in the United States to include filter strips along streams and generally encourage the retirement of land which has a significant negative impact on water quality could yield a budgetary saving of $1 billion per year from reduced commodity support payments. At the same time it is recognised that the programme is, at best, only an interim measure. In the longer run the United States Government considers that budgetary expenditure and agricultural supply can be more efficiently achieved by eliminating its price support programmes and removing price distortions in agricultural markets.

The European Community regulation on extensification also provides a framework for implementing schemes which are targeted to areas where the off-farm benefits of setting land aside from production are substantial. Examples of the possibility of doing this include setting land aside in groundwater recharge areas and setting land aside in strips along the edges of streams so that run-off is filtered and reduced. Most existing supply reduction programmes and also most of those which are currently under discussion, however, are not being developed in an integrated manner which seeks to simultaneously enhance environmental quality and reduce agricultural surpluses. For example, the 1987 Swedish surplus reduction programme which aimed to reduce crop and animal production by setting land aside for one year could have been targeted to farms and regions with surplus manure production and, also, soils sensitive to leaching. Instead, farmers were offered payments to set land aside as fallow, in a manner which increased nutrient leaching. As a result of this experience the current set-aside programme requires all land that is set aside to be sown to a cover crop. Another Swedish example is the introduction of incentives for forestry which do not take account of the relative sensitivity of different regions from a landscape amenity viewpoint. Incentives to convert crop land into forest land, for example, are being offered in forestry dominant regions where leaching is not a serious problem and the establishment of more pine plantations will only reduce landscape amenity values. To offset this problem the Land Management Act requires farmers to obtain permission before they plant arable land to trees, but the refusal of permission is rare.

As well as increased pollution the recent increases in and trends towards specialised crop production have resulted in losses in landscape amenity values, wildlife habitat and species diversity (Figure 5). In Germany, to contain this problem farmers are being offered payments to leave crop edges unsprayed and to leave headlands uncropped. In doing this a new role is being created for farmers as wardens of the landscape. The study concluded, however, that a significant political and public relations effort is needed if farmers are to accept that they would be as well employed accepting direct payments of this type, out of general tax revenue, rather producing more traditional agricultural products. In the long run, however, payment for such activities is likely to be more acceptable than payment to set land aside and do nothing.

Figure 5

LOSS OF HEDGEROWS IN GERMAN AGRICULTURAL LANDSCAPE DURING THE LAST CENTURY

1877

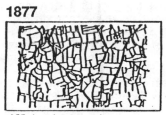

133,4 meters per ha

1954

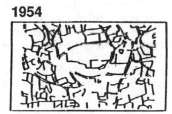

93,75 meters per ha

1971

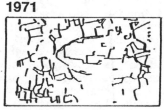

60,0 meters per ha

1975

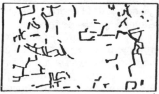

38,8 meters per ha

1979

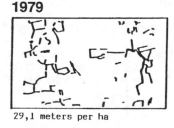

29,1 meters per ha

Source: MARXEN (1979), in N. KNAUER (1986)

41

In summary, the work in this policy area study indicates first, that the potential adverse public health and substantial environmental impacts are associated with the use of agricultural chemicals in intensive crop production and that many of these impacts are attributable to agricultural policies which increase production. Second, it is possible to integrate production limiting policies with environment protection policies. To date this has only been done to a very limited extent. Carefully designed supply reduction programmes could be designed to simultaneously reduce pollution, improve the landscape, provide wildlife habitat, maintain farm incomes and even reduce government expenditure through a shift towards direct income rather than commodity price support. But in doing this efforts will also need to be made to make farmers fully aware of the new role that they could have in providing and maintaining the environment. Third, input quotas and taxes offer a promising but as yet undeveloped means of reducing the cost of agricultural policies. Both high and low input taxes appear to be politically more acceptable when the revenue raised is used to finance agricultural pollution control measures. Both high and low taxes seem to have an additional psychological advantage of drawing the cost of chemical inputs to the attention of farmers and in some cases reducing the extent to which they use them at levels which are greater than the economic optimum. High input taxes could be used to reduce pollution, but need to be well in excess of 100 per cent if they are to significantly reduce agricultural pollution. Taxation of chemical inputs was found to be an efficient method of raising revenue for pollution control and prevention activities such as improved advisory services and research, but taxation levels high enough to bring about the desired reduction in chemical use might be difficult to implement and would probably reduce farm incomes to politically unacceptable levels. Fourth, other policy approaches including the targeting of land set-aside programmes to areas which are the source of environmental problems can produce both environmental gains and budgetary savings.

In conclusion and at a more general level, the details of successful approaches to controlling pollution from intensive crop production will vary considerably among different regions even within one country. Unless consciously planned and developed so as to take full account of regional disparities in physical and economic conditions, benefits to agriculture and the environment are unlikely to be realised from the above policies. Moreover, in the light of the problems associated with any individual measure, the most successful way to overcome these problems is likely to require a multiple approach. The set-aside of environmentally sensitive land which is important for water supplies, for example, could be coupled with transitionary compensation programmes for the requirements to adopt new cultural practices; the encouragement of diversification; and subsidies for farms to maintain or re-establish particularly attractive or valuable environments.

Intensive animal husbandry and the management of animal manure*

This policy area study focused on the increasingly serious problems of air, soil, and water pollution caused by intensive animal husbandry. Of all of these the pollution of water supplies is usually regarded as the most

---

*   This section summarises the ad hoc Group on Agriculture and Environment's studies on the impact of pollution from other sources on agriculture. The work is summarised in greater detail in chapter 6 of annex 2 to this report.

serious problem, as nitrates can form nitrites which are harmful to young children. In addition nitrites can react with other substances such as secondary amines to form carcinogenic nitrosamine compounds. Odour and visual damage to the appearance of the landscape are also problems (Table 3).

Table 3

POLLUTION FROM INTENSIVE ANIMAL HUSBANDRY:
FACTORS, EFFECTS AND CONSEQUENCES

Factors	Effects	Consequences
1. Siting of the buildings	Noise, appearance, smell	Neighbourhood and aesthetic problems
2. Run-off from spreading operations	Smell, infiltration Production of nitrates Oxygen absorption in flowing water, Leaching of organic compounds, nitrogen and phosphorus, Eutrophication of slow-running water and possibly estuarine and coastal water, Transport of pathogens Bacterial contamination of shellfish marketing	Neighbourhood and tourism problems Drinking water quality Disruption of river ecosystems Potable water quality Effects on human health and shellfish
3. Inadequate manure storage capacity	Spreading at sub-optimal times	All of above
4. Atmospheric emission of ammonia	Acidification of the environment through depositing and leaching	Toxic effect on plants and also the soil

Examples of the impacts of these problems are striking:

-- in 1980 only one area in Brittany had nitrate levels in excess of the European Community maximum permitted safe level for potable water supplies of 50mg of nitrate per litre, but by 1986 this number of areas had increased to 6 and in 21 other areas nitrate pollution had risen to above 40mg per litre;

-- similarly in The Netherlands, it is estimated that within the next few years it is likely that 25 per cent of current groundwater supplies will contain in excess of 50mg of nitrate per litre;

-- in The Netherlands 43,000 farms produce more manure than the intended maximum limit of 125kg of phosphate planned to apply from the beginning of the year 2000, representing an annual surplus of 18 million tonnes of animal manure; and

-- 20 per cent of acid deposition in The Netherlands now comes from the ammonia released by animal manure, with significant adverse effects on trees in some areas.

These problems are not confined to The Netherlands and France. Widespread examples of them can also be found in Belgium, Denmark, Germany, and Sweden. Local examples can also be found in countries like Switzerland. An important feature which emerges from examining the extent of pollution from intensive animal husbandry in all countries is the dominant influence of regional disparities in soil and climatic conditions. Much of Brittany, for example, is underlain by a furrowed precambian rock formation which prevents the lateral spread of nitrates to adjoining groundwater and limits the use of groundwater to local areas. Consequently in Brittany there a greater dependence on surface water supplies for drinking water while in The Netherlands, as the surface is flat and the soils sandy, most drinking water is drawn from underground sources. The problems also tend to be compounded by the varying degree to which mineral fertilizers are used in each country.

Many of the above impacts have been worsened by the growth and intensification of animal husbandry in recent years and its separation from crop and pasture production. In The Netherlands, for example, production of pork has increased by 87 per cent and egg production by 138 per cent since 1970. Many farmers purchase a substantial proportion of their feed and they often have insufficient land to spread the manure they produce in a manner which does not cause pollution.

Farmers also often appear to be unaware of the fertilizer value of the manure they produce. A survey of farmers in Brittany, for example, found that barely one third of farmers were aware of the fertilizer value of their manure and that, as a result, they often cause water pollution by over-applying mineral fertilizers. Surveys in other countries have come to similar conclusions. Consequently, in most OECD countries where nitrate pollution is a serious problem, agricultural advisory and extension services are being strengthened and targeted to problem areas. The principle theme of the advice is that by making better use of their animal manure farmers can simultaneously increase their income and reduce agricultural pollution.

In the countries where pollution from intensive animal husbandry is a severe problem there has also been a trend to making farmers pay the costs of reducing and controlling nitrate and phosphate pollution. In The Netherlands, for example, a general levy on manufactured feed has been introduced to make farmers contribute to the costs of conducting the research and providing the advisory services associated with this problem. In addition, a levy on surplus manure for farms which produce more than 125kg of phosphate per hectare per year has also been introduced to finance the establishment of a national manure bank and the construction of central manure storage and processing facilities. Feed suppliers are also being persuaded that they must, in their own interests, begin to construct such facilities.

Denmark has considered, but rejected, such a levy in favour of a series of land-use regulations which require farmers, at their expense, to adopt less polluting practices. Interestingly, in 1964 France introduced legislation enabling the state representatives (Préfets) to set water quality objectives and the basin water authorities to collect the cost of pollution control through a levy on farmers. This system, however, has not been put into operation in all water basins. It exists in the Seine-Normandie and Rhône-Mediterranée-Corse basins, but as yet has not been implemented in the Loire-Bretagne basin. The reason this levy has not been introduced in the Loire-Bretagne is that it has proved impossible to adequately establish a cause-effect relationship which enables responsibility to be partitioned among the basin's dairy, pig and vegetable farmers at a local level. Together, the above observations reveal the big difference between theory and practice. Implementing and enforcing regulations is an order of magnitude more difficult than creating them.

A significant difference between France and The Netherlands is the cost structure facing farmers. In France, pigs predominantly are fed on cereals at prices set by the Common Agricultural Policy while, in The Netherlands, a significant part of livestock feed, especially cassava and corn gluten meal is imported either duty free or at a low rate of duty. At the Community level there is an import quota on cassava and, as a result of the ability of Rotterdam to handle large bulk livestock feed carriers, Dutch farmers have a competitive advantage over other producers, because of the lower transport costs. As a result of this advantage and also a number of other factors such as entrepreneurship and technology, intensive animal husbandry has tended to concentrate in The Netherlands and the animal manure problem is greater there than in any other country. From this it is apparent that international trade and tariff arrangements can have an adverse influence on the location and intensity of agricultural production and, through this, the severity of pollution problems. Decreased support for cereal production or increased tariffs on cassava and other feed inputs, for example, could cause some Dutch intensive animal production to shift to other countries and regions and, through this induced structural change, reduce nitrate pollution.

Another problem is that manure storage facilities are not always adequate and that spreading and disposal practices often lead to run-off into water supplies. Typically, countries have responded to these problems by introducing planning controls and regulations which, for example, prohibit the spreading of manure on frozen ground and require farmers to be able to store their manure for at least 6 months. Usually farmers are required to meet the costs of these restrictions. Enforcement, however, is a problem and there are many examples of these restrictions not being enforced. The studies conducted by the ad hoc Group in this and other policy areas suggest that the effectiveness of regulations in achieving an integrated solution to pollution problems associated with agricultural production is heavily dependent on the degree to which they are enforced. During the transitionary period when stricter regulations have been first introduced, however, farmers in all countries have been offered some financial assistance. For example, in Denmark, farmers are being offered a 30 per cent subsidy to upgrade the manure storage facilities to 9 months storage capacity.

Obviously a mixture of measures is needed. On the one hand, it must be made possible for the Brittany farmers to install pollution control equipment, either through improving their cost structure or through direct financial assistance during the transition period while environmental standards are upgraded. In the interests of the integrated and more efficient use of resources, such assistance might be justified by the export subsidy savings accruing to the Community budget through the Breton farmers' use of surplus Community-produced feed. On the other hand, the volume of potential pollution must be limited, perhaps by controlling the number of livestock which can be kept per hectare of land available for manure spreading and the phosphates and nitrates used for crop and pasture production.

In The Netherlands, although there has been strict control of new entrants into the industry and on expansion of current facilities since 1984, livestock numbers have continued to expand. Progressively increasing restrictions on the amount of manure which may be spread per hectare are now being introduced (Table 4). These standards, depending on the circumstances, translate to a maximum of between 4 and 12 cows per hectare. Swedish experience, however, suggests that most crops can only remove less than 50 kg of phosphate per year and hence some people are expressing the view that the standards set will not completely arrest current phosphate and nitrate pollution problems. As an additional policy measure the current regulations require that any excess manure must be transferred, at the expense of the producer, from manure surplus to deficit areas.

Table 4

MAXIMUM PERMITTED APPLICATION OF PHOSPHATE FOR ANIMAL MANURE
IN KG PER HA PER YEAR IN THE NETHERLANDS

Phase	Period	Grassland	Fodder maize land[1]	Arable land
1	1 May 1987 - 1 Jan 1991	250	350	125
2	1 Jan 1991 - 1 Jan 1995	200	250	125
3	from 1 Jan 1995	approx. 175	approx.175	125
4	from approx. year 2000	final standard	final standard	final standard

1) Average, standards vary with soil type.

In addition to the above regulations, in Denmark farmers are being required to prepare management plans as a means to reduce nitrate and phosphate pollution. Similarly, Dutch farmers are required to keep records which show how they will dispose of the manure on their farms in a manner which meets the phosphate criteria set out in Table 4. In Denmark, however, farmers are required to prepare plans which give full consideration to the management of nitrate and phosphate from all sources. The advantage of these plans is that they not only make farmers think through the implications of what they are doing to the environment, but that they also significantly reduce and internalise enforcement costs.

In summary, the conclusions which arise from this policy area are: first, that by strengthening and targeting advisory services farmers can be persuaded to improve the efficiency of animal manure spreading and thereby simultaneously reduce farm costs and agricultural pollution. But unfortunately these measures are often not sufficient to overcome the problem. Second, in many areas, the introduction of regulations will also be necessary. Providing these regulations are enforced, they will be most efficient in helping resolve the problem if they are implemented in a manner which makes farmers meet the cost of pollution prevention and control. In developing and implementing and enforcing these regulations it is important to give careful consideration to the regional disparities which exist within and among countries. Financial assistance, however, may be necessary when stricter regulations are first introduced. Mandatory management plans can be used as one means of implementing these measures and overcoming the problems of identifying farmers whose agricultural practices cause pollution and of dealing with regional disparities. Third, levies and input taxes can be used as a means to internalise the costs of reducing and controlling agricultural pollution. Finally, in developing integrated agricultural and environmental policies consideration must be given to the effects of production support, international trade and tariff arrangements as they appear to have a significant effect on the location and intensity of agricultural production.

Dryland agriculture, soil conservation and the management of soil erosion

Soil erosion is a growing problem in many OECD countries and in recent years some have come to realise that the off-farm costs of soil erosion are generally greater than the on-farm costs. In the United States, for example, one study has estimated that the off-farm costs of soil erosion are in the vicinity of US$2-6 billion per annum which is roughly one-half to one order of magnitude greater than on-farm costs of soil productivity losses. Another study has estimated that off-farm damage accounts for 90 per cent of the total costs of soil erosion (Figure 6). A key environmental problem is that farmers only have an incentive to control on-farm costs and, unless forced to absorb the costs of off-farm damage from soil erosion, are usually unprepared to take action to reduce off-farm damage. It is recognised that the control of on-farm erosion will produce off-farm benefits, but in cases where for example the soil is deep, there is little short term incentive for farmers to control erosion. A second problem is that the damage is often distant from the source and often occurs several years after the erosion. A third problem is that it is often very difficult to identify which farmers are responsible for the damage on a particular site. Consequently, soil erosion and its prevention, soil conservation, is often perceived as a classic non-point source or diffuse pollution problem.

Soil erosion and soil conservation became issues in countries like Australia and the United States in the 1930s but, although it was recognised that off-farm sediment damage was an important problem, primacy was given to the threat of soil erosion to agricultural productivity. Even when agricultural impacts on the environment became an issue in the late 1960s, the main worry was the effect of pesticides on wildlife and human health and Departments of Agriculture in most countries continued to concentrate on maintaining soil productivity rather than reducing the off-farm effects of soil erosion. As a result of this approach, until 1985 farmer participation in erosion control programmes was largely voluntary, encouraged by cost sharing, technical assistance, education and persuasion. In several countries, a number of programmes have sought to stress that farmers are stewards responsible for the land they manage. Their role in caring for the land for future generations, especially their children, has also been stressed. With the exception of urban water catchments and land immediately adjacent to streams, land-use regulations have virtually never been used in OECD countries to reduce the off-farm impacts of soil erosion.

In the United States it is only since the early 1970s that environmental assessments and impact statements have usually been prepared during the formulation of farm price and income support policies. But even though this process began in the 1970s, no truly integrated policies emerged until 1985. Inducements to retire land in order to reduce surplus production, for example, were not directed to the land most at risk of soil erosion and the contracts were for one year only which meant that the conservation benefits from these set-aside programmes tended to be either minimal or negative. Moreover, as the benefits depended in part on the area of crop land, there was an incentive to cultivate land that otherwise would not have been cropped, as a method of increasing eligibility.

The 1985 Food Security Act provides for new environmentally oriented measures, including the set-aside of land to reduce agricultural surpluses in the conservation reserve programme and cross-compliance provisions which require conservation measures in return for eligibility for benefits under support programmes. Under the conservation reserve programme, farmers submit bids to take highly erodible land out of crop production for 10 years and place it in a conservation reserve which may be planted to trees, to grass or to wildlife cover, but not cropped or grazed. Farmers are paid a rental fee for their land. This is a clear example of the effective integration of agricultural and environmental policies. It is also a clear example of the environmental benefits which can be achieved by targeting programmes to problem areas. The benefits of such programmes would be greater if they were targeted so as to minimise off-farm, rather than on-farm soil erosion damage costs. One obvious way of doing this is to offer greater incentives to farmers who set land aside in filter strips along the edge of streams in a manner which reduces the run-off of soil, pesticides and minerals into the stream.

A second example of the recent emergence of integrated policies in the United States is the inclusion of the "swamp-buster" and "sod-buster" provisions in the 1985 Food Security Act. These provisions acknowledge the fact that production incentives encourage farmers to expand into more marginal areas and seek to prevent this from happening. This is done by providing that farmers who expand production into highly erodible land lose all their entitlements to participate in Federal Government farm price support and other financial programmes unless they implement an approved conservation plan which keeps erosion within an acceptable level. Farmers who drain wetlands and produce a crop on them are also excluded from farm price and other support programmes.

Figure 6

ANNUAL COST OF CROPLAND EROSION IN THE UNITED STATES
ON-SITE PRODUCTIVITY LOSS AND OFF-FARM DAMAGES
(1982 DOLLARS)

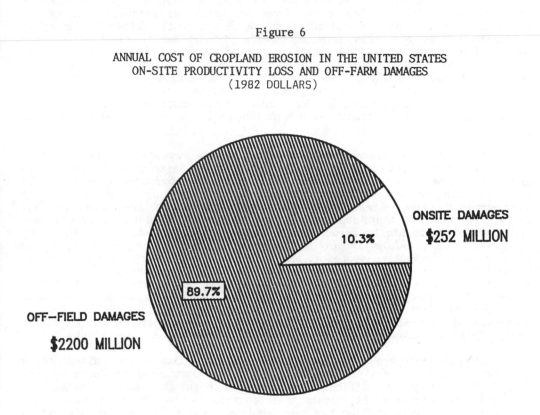

Source: Second RCA Appraisal

The cross-compliance provisions mean that after 1990, to receive crop deficiency payments and most other forms of financial assistance, farmers of highly erodible soils must prepare and have approved conservation plans which reduce erosion to an acceptable level. These plans must be fully implemented by 1995. A key feature of the above programmes is that, unlike the many European programmes which aim to increase production and farm income, most American programmes involve direct payments and loans to farmers which can easily be made conditional on other activities. In Europe, however, most assistance is offered to farmers via indirect market intervention policies which preclude the attachment of conditions to their payment.

In Portugal, the Soils Service considers that the present 51 per cent of land in crops ought to be reduced to the 28 per cent of land which is actually capable of sustaining grain production on a permanent basis. After soil erosion, the remaining shallow soils frequently become waterlogged in areas where the slope is less than 6 per cent and sediment deposition on crop land and in drainage channels sometimes becomes a major problem. In the case of steeper slopes the problem comes from too high an intensity of cereal production due to high, subsidised prices which are nearly double the European Community prices and also from Portugal's Mediterranean climate. Mention should also be made of certain social reasons as farmers have few alternatives. Over the next decade, however, these prices are to be reduced to European levels and, as a result, many farmers are no longer prepared to adopt conservation practices. Appropriate payments to encourage rapid extensification and diversification into other environmentally desirable activities during this transitionary period could be used to overcome this problem.

In an effort to reduce soil erosion problems Portugal offers subsidies and concessional loans which are easily available to farmers for reforestation and establishment of permanent pastures. Similar programmes also exist in Australia and the United States. A problem with this approach of offering inducements to farmers via loan, cost share and subsidy programmes is that farmers often decide that, unless they receive subsidy payments, they will not adopt the environmentally favourable management practices which are necessary to reduce the off-farm impact of their activities. The contrasting and seemingly more effective approach, which is being used by countries to reduce non-point source pollution from intensive animal husbandry, is to offer financial incentives only during the transitional period when more stringent land-use regulations are introduced to enforce regulations which internalise the off-farm damage prevention costs. Management plans, similar to the conservation plans which are being required for cross-compliance in the United States, are being used to assist in enforcing these land-use regulations.

Although there are more apparent inter-agency links in the United States the extent of real institutional integration as contrasted to "organisation chart" integration is difficult to judge in both the United States and Portugal. Nevertheless, it appears that progress in reducing soil erosion and encouraging soil conservation has been hindered by the lack of administrative arrangements which encourage the integration and that it is only in the last few years that efforts have been made to improve this situation. In recent years the introduction of administrative arrangements which facilitate and reward inter-agency staff exchange has been successfully used as a means to improve inter-agency collaboration.

The reasons why earlier attempts failed to link agricultural and environmental policies so that they acknowledge off-farm sediment damage costs and are targeted to areas where rates of erosion are high, include the lack of sufficient information on off-farm impacts and the lack of informal administrative arrangements which are necessary to facilitate the development of integrated policies. A feature of the progress which has been made in the United States is the impressive data collection, monitoring and modeling effort which was necessary to enable these integrated policies to be implemented. But data on benefits and costs of conservation measures is still lacking and this is considered an important area for research.

In summary, the conclusions which arise from this policy area study are first, that agricultural policies can be formulated and implemented in a manner which enhances their positive effects on the environment and reduces their adverse effects. The cross-compliance, sod-buster and conservation reserve provisions of the Food Security Act are specific examples of integrated policies which do this. Cross-compliance appears to be an extremely effective policy instrument for achieving integration and could be used to reduce pollution from pesticide and fertilizer use. Where subsidies exist and are paid directly to farmers it could also be used to make irrigation practices and land management practices such as the spreading of animal manure more environmentally favourable. Second, the development and implementation of integrated policies is dependent on inter-agency co-operation and goodwill. In the past efforts to develop integrated policies have often been hindered by interdepartmental jealousies and territorial disputes. Staff exchange is being used as one means of overcoming this problem. Third, the establishment and use of extensive monitoring and information systems is often a necessary precondition to the development and implementation of integrated polices. Fourth, the use of conservation and other such plans provide an efficient and effective means of reducing the administrative costs of ensuring that farmers do not adopt practices which cause environmental damage such as erosion. Fifth, in setting land aside from production it is important to provide a sufficiently long period for conservation benefits to be realised. Sixth, there are significant environmental advantages in targeting such agricultural policies to regions with environmental problems. Finally, it is apparent from the Portuguese experience with farmers in the face of falling prices that the continuation of environmentally favourable practices is unlikely to last during periods of financial stress. Incentives to encourage diversification and extensification can help overcome this problem. These incentives should encourage a change from cereal farming, particularly where there is erosion, to reforestation and to improved pastures.

Changing landscapes, land-use patterns and the character of rural landscapes*

This policy area study focused on the continuing role of farmers in maintaining agricultural landscapes amid changing circumstances and conditions. The clear message from the study is that farmers play an important role in maintaining and providing wildlife habitat, maintaining agricultural landscapes which are appreciated for their aesthetic, tourist and recreational value, and in many areas contribute to regional infrastructure and development. A key problem is that many, if not most, of the areas which are environmentally sensitive are marginal in an agricultural production sense. Thus, small changes in agricultural policy can have a big impact on the viability of agriculture in these regions.

---

* The two regions examined in the British policy area study were the Norfolk and Suffolk Broads and new conservation initiatives on farm estates in East Anglia and the North Pennine Dales. The Austrian example came from the Tyrol, in the rugged mountainous western region of the country. Both these studies are summarised in greater detail in chapter 5 of the annex 2 to this report.

There is also a more general concern about the rate of landscape change in many countries. This has led several European countries to introduce schemes to prevent the clearing of hedgerows and South Australia to ban the further clearing of native vegetation and offer farmers compensation for the lost production value on the condition that they accept a covenant on their land title which prevents further clearing. Figure 7 provides data on the rate of landscape loss in England and Wales.

Figure 7

LOSS OF AND DAMAGE TO WILDLIFE HABITAT IN ENGLAND AND WALES SINCE 1945

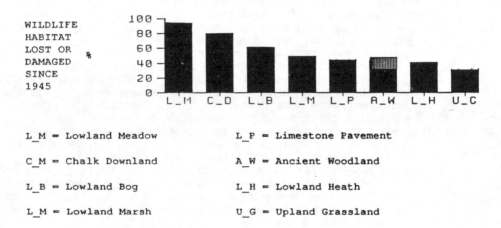

WILDLIFE HABITAT LOST OR DAMAGED SINCE 1945 %

L_M = Lowland Meadow	L_P = Limestone Pavement
C_M = Chalk Downland	A_W = Ancient Woodland
L_B = Lowland Bog	L_H = Lowland Heath
L_M = Lowland Marsh	U_G = Upland Grassland

Source: Adams, W. "Nature's Place", Allen & Unwin, 1986.

One British example of an environmentally sensitive agricultural area where landscape change is an issue is the Norfolk and Suffolk Broads area in East Anglia where reed fen, fen woodland and drained grazing marshes, which are renowned for their wildlife value, recreational potential and traditional open grazing marsh landscapes are declining. Here the problem is that the farmers who use the unfertilised marshes are experiencing recession due to the rising costs of maintaining adequate drainage and falling livestock prices. The cropping of these Broads is also being encouraged by the high prices offered for cereals under the Common Agricultural Policy. As either a move to cropping or the abandonment of the draining would destroy most of the recreational and landscape amenity potential of the area, financial incentives in the form of five year agreements are being offered for farmers to continue to manage the Broads in a traditional manner. No permanent agreements, however, have been made to preserve this landscape. Legislation is currently under discussion which would create a new Broads authority whose duties would include conserving and enhancing the natural beauty of the Broads.

Another example is the Pennine Dales in the United Kingdom which contains an outstanding area of habitat interest and landscape beauty in the limestone hills of North Yorkshire and South Northumberland. The region is almost wholly devoted to livestock grazing, notably sheep. The ecological distinctiveness of the Pennine hay meadows is dependent upon low nutrient status, well-timed grazing and hay removal to preserve its flower-rich hay meadows. The traditional stone barns and walls that characterise this very special landscape are also of interest.

A third example is the Tyrolean area in Austria where the quality of the farmed rural landscape has a direct bearing on the viability of the tourist industry. Under a Special Programme for Mountain Areas the Austrian government provides payments to preserve areas that are in danger of being abandoned. Most of these funds are allocated to direct income support, but special assistance is also offered in particular regions and for land levelling, forest management and the construction of utility lines. Interestingly, in addition to this, the local tourist boards are financing and providing financial incentives to farmers to retain traditional practices which attract tourists. These mostly involve hay mowing allowances so that meadowlands are cut to maintain the visual appearance of cut grass fields and the condition of ski slopes important for the tourist industry.

Two main policy tools are being used to protect the landscape and habitat of the Pennine Dales. The key meadow habitats have been granted conservation status as Sites of Special Scientific Interest which are protected by legislation that largely ensures that farmers cannot alter their ecological or geological characteristics. This legislation provides for the payment of compensation in cases where this leads to a loss of agricultural income. In addition, all the important Dales have now, like the Broads, been designated as Environmentally Sensitive Areas in which farmers are offered financial incentives to maintain traditional patterns and practices. A weakness in the financial incentive scheme is that most of the agreements with farmers only run for 5 years and hence there is no guarantee that the area will continue to be protected.

In all the above examples it is apparent that the aim of the authorities is to guide possible future agricultural practice so that it enhances or at least maintains the biological and landscape diversity of the area. This means steering future management into patterns of use that are traditional and unintensive. They involve maintaining livestock regimes where arable farming is more profitable than extensive livestock grazing (the Broads) or low fertilisation grassland management when intensive grass production would be more worthwhile (the Dales), or livestock regimes where farming would otherwise barely exceed subsistence levels (the Tyrol). But in every case agriculture is essential if the environmental value of these areas is to be retained.

The policy measures being considered and implemented by governments to maintain farming in environmentally important areas involve:

i) Recognising and guiding socio-economic changes in marginal agriculture areas by encouraging farm income diversification and promoting tourism;

ii) Providing direct income support for small, marginal farmers who cannot benefit from any other income diversification trends, but whose presence and practices serve vital social, economic and environmental functions;

iii) Financing special management measures, including retaining the status quo in landscape features and habitat ecology by means of special agreements to area-based payments;

iv) Facilitating indirect income support through capital or recurrent grant aid for farm restructuring or socio-economic adjustments;

v) Protecting prime environmental habitats and landscape areas by means of special legislation, landscape designation, planning and development controls, or enabling land ownership to be placed in the hands of accredited conservation agencies;

vi) Creating a network of supporting advice, information, interpretation and education to assist farmers to include a conservation perspective amongst their farming activities; and

vii) Devising means for the diversion or abandonment of surplus agricultural land by special measures, so as to encourage the creation of new conservation in farmed areas, or the establishment of new wild areas on redundant farms.

These measures overlap but are essentially complementary and, in each case, seek to create a new role for farmers in managing the environment. Their success will depend upon the particular environmental, historical and economic characteristics of individual farms and rural communities.

Major conservation sites, regarded as having international significance for their landscape typicality, ecological rarity and genetic uniqueness, will always require special protection and particular management. Often either government acquisition or the transfer of ownership to conservation agencies who may establish special lease arrangements with farmers is required in such circumstances.

Where there is a desire to preserve established agricultural landscapes, a mix of land-use restrictions, management agreements and income support measures may be used. Income support may be suitable in zones of special socio-economic hardship where alternative economic opportunities are limited, but in most cases it would be more efficient to couple this support with a requirement to produce environmental benefits. Management agreements involving special payments are desirable where production incentives are causing environmentally damaging farm intensification. The amount of supplementary financing needed will usually depend on the relative support for "intensive" agriculture available through public subsidy, and the opportunities for non-farm income generation. A reduction in production incentives will generally make it possible to reduce the need to offer financial incentives.

Opportunities for promoting local and regional initiatives also exist. Peer group pressure and neighbourly encouragement can also have a profound educative effect in encouraging farmer compliance with voluntary measures. This is particularly true when the schemes are locally administered by farmer representatives. Much can be achieved by encouraging clusters of neighbouring farms to share their ecological and landscape "capital" and benefit from joint marketing ventures.

54

The more general conclusions from the work in this policy area study are first, that the reduction of price support could adversely effect marginal production areas with highly prized habitats and landscape features. It is recognised, however, that these areas usually account for a small proportion of total production and that consequently the above issues should not be used as a reason to maintain production incentives. Second, declining farm incomes will often cause farmers to change the management practices which they adopt. These changes, especially during the adjustment period, can have substantial environmental repercussions in marginal agricultural areas where environmental quality is dependent on the retention of traditional agricultural practices. Third, the adverse effects can be overcome by providing financial incentives for the retention of activities necessary to maintain environmental quality. Policies orientated so as to promote agricultural diversity tend to benefit the environment by contributing to ecological diversity. Fourth, a clear distinction needs to be made between schemes designed to preserve environmental quality and others such as those with social welfare objectives. It is often possible to design schemes to achieve both social welfare and environmental quality objectives.

4. IMPROVING POLICY INTEGRATION

This section sets out some concepts for successful policy integration associated with administrative arrangements and the use of policy instruments and, in particular, considers the relative merits of using the advisory, economic and regulatory approaches outlined in sections 3.1 and 3.2 of this Report within the general context of the ongoing process of agricultural policy reform and the application of the Polluter-Pays Principle as agreed by the Governments of OECD countries.

Integration requires policy makers to give full consideration and to accept responsibility for the effects of their policies on the objectives of all other sectors. This is as true for the effects of environmental policies on agriculture as it is for the impact of agricultural policies on the environment. One necessary precondition to this concept of "responsibility" is that the relevant policy-makers must jointly agree to the objectives of interdependent and interacting sectors and recognise and accept the nature of the trade-offs to be made among these objectives where they compete or conflict. Greater progress will be made if objectives are clear, measurable and time specific.

Successful integration also presupposes an administrative structure which fosters the integrated selection and implementation of a coherent set of instruments which, by making appropriate trade-offs, will jointly achieve these objectives. Coherent strategies for agriculture and the environment, such as those being developed in countries like The Netherlands which involve three departments in the preparation of four year plans for agriculture and the environment, have a major role to play in resolving conflicts between agricultural and environmental policies.

Institutional integration

The administrative arrangements which are best for the development and implementation of agricultural and environmental policies will vary from country to country, reflecting national attributes and circumstances. But it is clear that administrative arrangements exert considerable influence and, in many countries, are impeding the development of integrated policies. If better integration is to be achieved then two principal matters require attention.

First, the objective of policy integration per se must be given explicit recognition. Political support and a public commitment to achieving integration must be obtained from ministers of agriculture, ministers of environment and others who are in a position to arbitrate inter-departmental disputes. Mandatory consideration of environmental issues in any policy and legislative proposals will obviously aid the process of integration. In many cases this need for the mandatory consideration of environmental consideration requires significant changes to the administrative procedures followed during policy formulation.

Second, in the light of existing integrated policy needs, existing organisational structures, notably inter-departmental arrangements for consultation, collaboration and communication, must be reviewed with regard to the management tasks and policies to be implemented. The OECD's work on the integrated management of natural resources indicates that the reality of divided jurisdiction, shared responsibility, overlap and even duplication at the boundaries of programmes needs to be recognised.

Integration procedures

Integration requires that policy instruments designed to achieve particular objectives in one sector are subjected to prior assessment for their effects on other sectors. For example, in introducing an environmental policy to control non-point sources of pollution, it is necessary to consider the effects of any restrictions on farm income. Similarly, policies designed to maintain farm income must be examined for their effects on the environment. Thus, in order to develop integrated policies, it is necessary to develop administrative procedures which ensure that the effects of one sector's policies are evaluated in terms of their likely effects on all related sectors.

Some of the successful approaches adopted by countries are formal, while others are informal and rely heavily on the goodwill of the agencies involved. Formal procedures, especially those which are backed by legislation appear to be successful. Examples of these formal procedures include mandatory requirements for inter-departmental consultation and public participation and requirements for the formal preparation of an environmental assessment of a proposed policy.

Integration instruments

Voluntary or advisory approaches, albeit backed by regulation, are extensively used in most OECD countries to achieve agricultural objectives. Successful agricultural advisory approaches take full account of all the economic conditions faced by farmers. They recognise the interests of the farmer while seeking above all to persuasively provide additional information about the benefits of adopting new practices.

Problems, however, have been encountered in using only advisory methods to persuade farmers to adopt environmentally-favourable practices. Consequently, in most countries, advisory approaches are supplemented by regulations and economic incentives. Incentives are used to change the economic behaviour of farmers, whereas regulations, usually in the form of accepted minimum standards of practice, are used as a last resort and where there are distinct externality issues which can be addressed through regulations.

An approach being tried in a few countries and being discussed in others is the use of input levies and quotas. Experience, particularly with water resources and, more recently with fertilizer levies, has revealed that both these instruments appear to be effective in increasing farmer awareness of the consequences of changing economic conditions and over-application of inputs.

Input levies and quotas can be effective in helping to reduce agricultural pollution. Increasing the cost of agricultural inputs makes some farmers more aware of the additional profits which can be made by the more efficient use of inputs and reduces both on- and off-farm pollution by internalising some of the off-site costs of pollution from agriculture. On the other side of the economic equation, price policy can be and is used to influence and, in particular, to change the intensity and location of agricultural production.

If prices are temporarily high or expected to fall in the near future, then in the absence of enforced sanctions, farmers may not perceive it as being in their interest to adopt or maintain non-polluting agricultural practices. In such cases, when future conditions are uncertain, the maintenance of environmental quality may require the strict application and enforcement of regulations.

Regulations, backed by appropriate sanctions, should also be used when the consequences of an undesirable environmental practice are morally unacceptable. The dumping of poisonous chemicals in a stream is clearly an unacceptable practice. Similarly, the need for a pesticide licensing system to control cancer and other human health risks is now accepted in all OECD countries. Clearly, property rights should not confer on farmers the right to pollute or reduce the value of off-farm resources.

When the issue is one of maintaining or enhancing the environment, incentive schemes are proving to be particularly effective in modifying non-polluting agricultural practices to bolster the positive contribution which farmers make to the environment. Under these programmes, farmers volunteer to accept payments in return for a commitment to undertake landscape amenity and nature conservation practices such as the grazing of cattle on hillsides to ensure the survival of a butterfly population or the maintenance of a terrace or stone wall. Importantly, experience is revealing that the administrative costs of incentive-based approaches are often significantly lower than regulatory schemes, and most administrators have a strong preference for this type of approach.

If all externalities of agricultural policies could be internalised and all changes in relative prices anticipated, near-perfect incentive systems could be designed and a regulatory approach would not be needed. As indicated in section 3, however, it is impossible to anticipate changes in input costs, account for all externalities and account for all the local effects of a national incentive-based system. Moreover, the information costs associated with implementing such schemes are prohibitive and regulations, backed by appropriately enforced sanctions, can be more precisely targeted than economic incentives. In particular, they do not require knowledge of the nature of damage functions and are not sensitive to price variations. Consequently, although preference may be given to a voluntary approach, those practices which are necessary to maintain the minimum level of environmental quality to protect human health, etc., should continue to be defined by regulation.

Most regulations take the form of standards or constraints and define limits for what is acceptable or tolerable. Within these limits, it is usual to use advisory and voluntary incentive programmes to achieve the desired state of the environment, the economy and social well-being.

The question which remains is: should different principles apply to pollution control and prevention programmes on the one hand, and programmes designed to enhance the positive contribution which farmers can make to the environment on the other?

In 1972, all Governments of Member countries of the OECD adopted the Polluter-Pays Principle which provides that:

"The principle to be used for allocating costs of pollution prevention and control measures to encourage rational use of scarce environmental resources and to avoid distortions in international trade and investment is the so-called 'Polluter-Pays Principle'. The Principle means that the polluter should bear the expenses of carrying out the above mentioned measures decided by public authorities to ensure that the environment is in an acceptable state. In other words, the cost of these measures should be reflected in the cost of goods and services which cause pollution in production and/or consumption. Such measures should not be accompanied by subsidies that would create significant distortions in international trade and investment." [OECD, C(72)128]

This Principle, which was reaffirmed in 1974, goes on to state that:

"This Principle should be an objective of Member countries; however, there may be exceptions or special arrangements, particularly for transitional periods, provided that they do not lead to significant distortions in international trade and investment." [OECD, C(72)128]

The rationale for this agreement is soundly based on the economic advantages of internalising pollution control and prevention costs. The emphasis is on allocating the costs of pollution control and prevention measures and, importantly, an exception is made in the case of the transition period when new control and prevention measures are introduced. In particular, the Principle does not justify the introduction or maintenance of subsidies which create significant distortions in international trade and investment. In several cases, compliance with the agreement will require countries to modify current policies.

In applying the Principle, it is important to determine who is the polluter. In the case of soil pollution by a fertilizer, for example, industry may be responsible for meeting the costs of controlling cadmium pollution and farmers for the costs of controlling nitrate pollution. In this case both agriculture and industry are responsible for accepting the cost of the control programme and passing these costs on so that they are ultimately expressed in the cost of consumer products.

It is recognised that difficulties are being experienced by some countries with the application of the Polluter-Pays Principle to agriculture but, as pointed out in sections 3.1 and 3.2, other countries have found ways to overcome some of these problems. Most of these solutions involve ensuring that at least the cost of any restrictions on agricultural practices are met by farmers and that, after any transition period, no subsidies are offered to reduce the cost of on-farm pollution control and prevention. Most of the problems which remain in applying the Polluter-Pays Principle are associated with identifying the polluter, finding cost-effective methods of enforcing the Principle and finding equitable methods of allocating the costs of off-farm control measures such as water purification.

In this latter case the consensus which appears to be emerging from countries which are trying to apply the Polluter-Pays Principle to agriculture is that input charges and levies should be used to internalise the costs of pollution prevention and control. This allocates the cost of pollution prevention and control to the users of the inputs and practices which cause the pollution and creates incentives for these users to stop causing the pollution. That is, when the difficulties in applying the Polluter-Pays Principle to agriculture at a local level cannot be overcome, then one alternative option is to collectively make the users of the practices and inputs which cause the problem, pay the costs of the pollution prevention and control measures. Subsidisation by taxpayers for pollution prevention and control should be abolished unless special reasons exist for its retention.

With this focus on the costs of pollution prevention and control it is necessary to answer the question: "What are pollution prevention and control activities?". This requires that a distinction be made between potentially polluting agricultural practices and those which provide positive environmental benefits. There is a continuum:

-- the need for adequate animal manure storage to reduce ammonia emissions, and through this acid precipitation, is clearly a case of controlling pollution;

-- while the removal of a hedgerow or woodland valued for wildlife and landscape amenity reasons is probably not pollution.

In many cases the distinction between polluting and non-polluting activities will be difficult to determine and, to some extent, must involve a policy choice at the national level. It would seem however that:

i) The Polluter-Pays Principle should apply to all agricultural policies and programmes which are designed to prevent, control or reduce both point and non-point sources of pollution;

ii) Financial assistance may be paid during a transition period when:

a) under a new programme, a farmer's environmental obligations are redefined; and

b) the payment of financial assistance would lead to more speedy environmental improvement.

As the need for financial assistance, however, is mainly a question of the need for speedy change, the decision of whether or not to pay it should be decided independently by Member countries. Moreover, such financial assistance should only be available for a pre-determined period and following this transitional period farmers, like industry, should be compelled to continue to meet their redefined environmental obligations at their own expense.

The Polluter-Pays Principle, however, can come into conflict with policies principally designed to meet other objectives. Consequently, it is proposed that, in the interests of achieving the full integration of agricultural and environmental policies:

iii) Payments made principally to achieve non-environmental objectives, but which have the effect of enhancing environmental values, should not be regarded as payments for pollution control or prevention. Such payments might include the provision of income support; the promotion of regional development; or the reduction of surpluses.

Table 5 provides examples of government programmes which comply with each of the above criteria which pertain to the implementation of the Polluter-Pays Principle and the exceptions which apply to it.

Table 5

EXAMPLES OF POLICIES IN OECD COUNTRIES WHICH COMPLY WITH
THE POLLUTER-PAYS PRINCIPLE, THE TRANSITION EXCEPTION AND
THE PROPOSED OTHER PROGRAMMES EXCEPTION

Criteria	Example
1. PPP applies to all	Dutch tax on the production of excess manure and a requirement that farmers dispose of all surpluses at their expense
2. Financial assistance	Swedish grants to farmers to upgrade manure storage facilities only during transition
3. Other programmes	The cross-compliance provisions of the Food Security Act in the United States which require farmers to adopt approved conservation plans in order to remain eligible for Federal agricultural programme financial benefits if they farm highly erodible lands.

In the cases where the question is not one of pollution control or prevention, but rather one of finding a mechanism to internalise the benefits to be obtained from redirecting agricultural activities to enhance conservation and other environmental values there is a need to offset those costs which discourage conservation. Management agreements provide one mechanism for doing this, others include the provision of economic incentives; the acquisition and conditional leaseback of a farm; the creation of conservation easements and strengthened environmental legislation; and the creation of mandatory requirements for farmers to adopt certain practices.

In the case of management agreements and other similar arrangements designed to compensate a farmer for the cost of providing a positive environmental benefit which requires investments and the adoption of practices in addition to those necessary for normal non-polluting agriculture, a distinction needs to be drawn between a) reimbursement for additional maintenance costs and expenditure; and b) compensation for lost production opportunities. In this latter situation:

 i) When a government wishes farmers to modify normal non-polluting land use and other practices to enhance environmental values, payment should be limited to compensation for the expected reduction in net farm revenue.

 ii) Payment should be conditional upon the establishment of legal arrangements such as conservation easements and restrictive covenants which guarantee that the desired public benefits will be realised;

iii) Programmes should be designed to preclude the possibility of farmers using the threat of anticipated environmental damage as a lever to obtain compensation.

 iv) Restrictions on agricultural practices which have a negligible effect on net farm income, such as the prohibition of the application of insecticides while a crop is in flower, should be implemented via mandatory requirements and not be compensated.

Table 6 provides examples of government programmes which comply with each of the above criteria, which pertain to the encouragement of agricultural practices and which enhance the positive contribution of agriculture to the environment.

Table 6

EXAMPLE OF PROGRAMMES WITHIN OECD COUNTRIES WHICH ENHANCE
THE POSITIVE CONTRIBUTION OF AGRICULTURE TO THE ENVIRONMENT

Criteria	Example
1. Compensation for lost	Payment of compensation to farmers on production value the Sommerset Levels ESA in the UK in return for an agreement not to crop them
2. Conditional upon guaranteed	When permission to clear vegetation is benefits refused in South Australia compensation is only paid when farmers agree to the attachment of a conservation conservation easement to their land title
3. Removal of incentive to threaten unintended action further clearing in South Australia	Compensation limited to the attachment of a permanent easement prohibiting
4. No compensation for minor restrictions	Restrictions on use of insecticides while crops are in flower in Belgium

5. EXISTING OPPORTUNITIES FOR THE SUCCESSFUL INTEGRATION OF AGRICULTURAL AND ENVIRONMENTAL POLICIES

In June 1985, Governments of OECD Member countries declared "that they will: "Ensure that environmental policies are taken fully into account at an early stage in the development and the implementation of economic and other policies...in such areas as agriculture..." and..."will promote the effective integration of these policies" (OECD, 1986). Within the context of agriculture and environment, opportunities for integration can be divided into existing and emerging opportunities.

Emerging opportunities for the integration of agricultural and environmental policies can be achieved by reforming agricultural policies to resolve existing surplus and trade problems. It is recognised that many existing opportunities are not dependent on changes to present-day agricultural policies and could be implemented immediately. Section 5 of this report describes existing opportunities which could be exploited immediately, and section 6 those which could emerge in the context of changes to agricultural policies currently under discussion. In the latter section, a plea is made to exploit, as far as is possible, the environmental opportunities which are being presented by the agricultural reforms under discussion.

Existing opportunities for improving the integration of agricultural and environmental policy, which do not rely on changes to agricultural policy include: opportunities for improving institutional arrangements; strengthening regulations associated with the use of agricultural inputs; enhancing advisory and extension services; changing regulations concerning product standards; removing impediments to adoption of environmentally-desirable practices; and strengthening monitoring and research activities. Some of these existing opportunities can be implemented at an individual farm level with little or no reduction in farm income and others present opportunities for farmers to increase the income they receive simultaneously, to reduce pollution and to enhance the quality of rural landscapes.

Institutional

In fostering integration, Governments need to create a political and administrative climate which is conducive to integration. Drawing on the experiences of all Member countries described in this report, it would appear that effective methods for achieving better integration include:

-- expressing, at ministerial level, a publicly announced commitment to integrating agricultural and environmental policies and ensuring that both types of policy do not have unintended negative effects one on the other;

-- promoting the development of agreed agricultural and environmental policy objectives which are clear, concise, measurable, time specific and, where appropriate, expressed in legislation;

-- encouraging an anticipatory planned approach to the development of agricultural policies which affect the environment and environmental policies which affect agriculture;

-- revising contradictory policies, such as the provision of subsidies and price support which encourage farmers to develop certain wetlands on the one hand, while providing different grants to those who agree not to develop the same or similar areas on the other;

-- facilitating and encouraging public participation in agricultural decision-making to ensure that full account is taken of the environmental consequences of implementing a proposed agricultural policy; and

-- encouraging the application of environmental impact assessment procedures to the development of agricultural policies and major projects.

Environmental responsibilities are allocated in many different ways in OECD countries according to their varied institutional, cultural, historical and economic backgrounds. There is a trend towards the creation of environment departments in many countries. In establishing and widening the responsibilities of these environment departments it should, however, be made clear to all other policy makers that they continue to be responsible and accountable for the effects of their policies on the environment. Environment policy-makers and departments are responsible and accountable for the effects of their policies on agriculture and, conversely, agricultural policy-makers and departments are equally responsible for the effects of their policies on the environment.

As, in larger countries such as France and the United States, the opportunities for integration are greater at a regional level, there appears to be a general need and more scope for a diversified regional approach to the development and implementation of agricultural policy than has existed in the past. In these larger countries homogeneous national and international policies, which are implemented on a sub-sector basis, have very different regional effects and, as a result, often have unintended negative effects on the environment. For example, support prices that barely maintain upland farming can lead to over-intensive farming in the lowlands. A stronger regional approach, however, must be backed by the development of a broad national perspective which facilitates regional decision-making but respects national and international obligations. Truly effective integration is more easily achieved when responsibility for the development and implementation of agricultural and environmental programmes is made at the same level.

If regional administrators can be left to account for the special circumstances of agriculture in the context of each local environment, then many administrative problems which so often preoccupy central administrations can be resolved at a level at which the remaining issues present clear choices. The promotion and facilitation of regional and local administrative autonomy can also be expected to enhance opportunities to maximise local and regional competitive advantages; to develop peer group pressure; and encourage people to develop a sense of regional responsibility. The planning process can also be strengthened at this level where opportunities exist for the integration of agricultural and environmental policies with all the other policies of a region.

As a result of the changing perception of the role of farmers in nature conservation, in some countries farmers are starting to develop their own conservation and tourist development schemes. Through this approach, they are increasingly becoming involved with local tourist boards and other development authorities. Government participation in farmer and regional conservation schemes is seen as being much more successful than farmer participation in government schemes.

Inputs

Agricultural inputs such as fertilizers, pesticides, feed additives and irrigation water have been responsible for many of the recent gains in agricultural productivity, but unfortunately a number also have had or threaten to have adverse side effects on the environment. Approaches to the control of the adverse effects of agricultural inputs vary significantly among countries but, in most cases, product quality standards for these inputs have been introduced to reduce the risks of immediate and, more seriously, long-term cumulative negative effects on the environment.

In the case of pesticides, all OECD countries now have mandatory registration programmes, but few regulations which restrict the quantity of pesticide which may be used. Pesticides banned in one country are still used in others. Opportunities exist for the further harmonisation of standards, registration procedures and testing methodologies among countries and can be expected to lead to environmental improvement through the more thorough evaluation of agricultural chemicals and the earlier detection of problems. Lower environmental quality may result, however, in that the least stringent standards may be adopted in the interests of harmonization.

In recent years the strengthening of water quality standards and the development of improved analytical methods has led to the detection of pesticide residues in ground water. In response many countries have recently either introduced for the first time, rewritten or substantially amended their pesticide legislation, reflecting rising concern about the impact of pesticides on the environment. In countries where pesticide legislation has not been reviewed in the last 10 years a need for substantial review may exist. Many recent reviews have been promoted by the detection of residues in ground water and data on the effects of pesticides on wildlife. Consequently, in countries where legislation has not been reviewed, there may also be a need to improve the monitoring of both ground water quality and species diversity, but this should not be used as an excuse for inaction.

Farmers tend to use agricultural inputs in response to economic factors. One of the important functions of advisory services is to encourage farmers to use only the quantity of fertilizers which is necessary for profit maximisation. In some countries there is mounting evidence that stable product prices encourage farmers to over apply fertilizers and that by using less they could increase profits and reduce water pollution. Such opportunities for increased farm viability and environmental improvement need to be realised. Countries have found that in areas where agricultural pollution is caused by the over-application of inputs, advisory and extension services can be used to increase farm income and reduce agricultural pollution.

Advisory services are usually free of charge but in several countries farmers have to meet either all or part of their costs. To encourage farmers in northern France to apply only the amount of fertilizer which is necessary for profit maximisation, free soil tests are being offered on an introductory basis for one or 2 years. It is hoped that once farmers appreciate the value of such tests, they will continue to test their soils once charges are reintroduced. Exploratory approaches which encourage farmers to adopt such environmentally-favourable practices should be encouraged.

Economic charges, particularly input levies, are being used to simultaneously reduce the over-application of inputs; to encourage the growth of alternative crops; and finance research. Opportunities exist for the greater use of input charges to reduce agricultural pollution and for the introduction of input quotas for fertilizers and pesticides. Opportunities also exist for the introduction of quotas on the inputs which cause agricultural pollution, but as yet no country has introduced quotas for this purpose. Water rights, which in many ways are a form of input quota, are being made transferable and, in countries where transferable water rights have been introduced significant reductions in water pollution have been observed. In countries where salinity from the over-application of irrigation water has become a problem marginal cost pricing for water resources is being introduced. In each case, the experience has been environmentally beneficial. Jointly, these experiences suggest that in areas where pollution caused by the over-use of inputs exists, by setting input pricing policies so that they cover marginal costs, including all external costs, and by introducing transferable input quotas substantial environmental improvement can be expected. These levies will be more politically acceptable and probably more environmentally effective if the resultant revenue is used to finance programmes which benefit agriculture.

Practices

Many agricultural practices make a very positive, and often essential, contribution to the maintenance of the environment. This, however, is not always the case. Traditionally, most of the concern about agricultural practices has been associated with the application of inputs and soil preparation. Operator safety and impacts on human health have also been of considerable concern. More recently, concern has expanded to consider the impact of these practices on wildlife, landscapes and the quality of natural resources. Intensive animal husbandry practices has also become an issue.

Dangerous practices, such as the inappropriate disposal of pesticide waste, are banned in most countries and, more recently, several have begun to introduce schemes which require all purchasers of farm pesticides to be licensed. Normally to obtain one of these licences operators must pass a course on pesticide safety. These new licensing programmes educate farmers about the impact of pesticides on the environment, the way their over-use affects farm income, and the merits of integrated pest management programmes. It is still too early to evaluate the effectiveness of such schemes.

In encouraging farmers to take greater account of environmental objectives and to use environmentally-favourable agricultural practices, most countries have a strong preference for using advisory approaches. Several of these advisory approaches have a strong focus on developing a farmer's sense of individual responsibility. Opportunities for enhancing this sense of responsibility include involving community leaders in the development of advisory programmes, running publicity campaigns which increase public pressure on farmers to improve the environmental aspects of their programmes and simply strengthening and enforcing regulations.

Many advisory programmes are made available to farmers on an equal basis but in recent years experience has revealed that substantial environmental gains can be achieved by targeting these advisory services to problem areas. A complementary approach, in the United States, has been to identify a range of "best management practices" for each targeted region and type of land use. This approach of defining best management practices and implementing them in association with other programmes has the triple advantages of a) forcing the local representatives of the different agencies formally to attempt to develop an integrated advisory package; b) presenting an integrated advisory package to the farmer; and c) requiring farmers who wish to participate in other agricultural programmes to adopt such best management practices.

In areas where agricultural pollution is not a problem and opportunities exist for enhancing the contribution which farmers make to the environment, several countries have begun to enter into management agreements with farmers. These management agreements with individual farmers foster close consideration of individual preferences and opportunities, within the regional context of the situation. Management agreements appear to be most successful when ongoing expenditure or farming activity is necessary to maintain environmental quality. However, they should be drafted in a manner that ensures that the expected improvement in environmental quality is attained. Cost-sharing programmes which subsidise certain capital improvements, such as the repair of a stone wall in preference to its replacement with a fence, provide another method of enhancing the environment. An alternative approach which is being used in several countries is to identify environmentally-sensitive areas and restrict land-use practices within them. This approach has been particularly effective in water catchments for reservoirs, but also has an application in areas valued for tourism, recreation and conservation.

More recently, attention in a number of countries has been focused on the need to remove economic and legal disincentives to the adoption of environmentally-favourable agricultural practices. In some countries for example, farmers are not permitted to use integrated pest management schemes if they wish to qualify for certain types of financial assistance. Impediments to the adoption of environmentally-favourable practices in agricultural programmes should be removed from all government programmes. There have also been a number of suggestions that serious consideration should be given to the introduction of pest damage insurance schemes which are applicable to integrated pest management and which reduce farmers' fears of crop yield loss caused by unusual pest outbreaks. As yet, however, no country has been able to overcome some of the pragmatic problems associated with the introduction of such a scheme.

In recent years several European countries have introduced restrictions on agricultural practices to control water pollution from non-point sources. These restrictions include the prohibition of manure spreading on frozen ground, limits on the quantity of manure which may be spread per hectare, requirements for mandatory manure storage capacity and restrictions on feed additives. Farmers are also being required to prepare spreading plans and ensure that a certain proportion of their land is sown with a green crop in the autumn.

Generally, all pollution control measures should be implemented at farmer expense in a manner which is consistent with the Polluter-Pays Principle. Subsidies should only be offered during the transitional period when storage facilities, etc. must be upgraded. In a number of countries, however, in apparent violation of the Polluter-Pays Principle, farmers are being offered subsidies to adopt environmentally-desirable agricultural practices which reduce non-point pollution such as not cropping next to water courses, adopting minimum tillage practices, building contour banks and not using fertilizers in water catchment areas.

Governments of Member countries adopted the Polluter-Pays Principle for reasons of economic efficiency and the expected benefits from partially internalising the costs of pollution. In each of these non-point source pollution reduction programmes the integration of environmental and agricultural policies would be more efficiently achieved by converting subsidies to transitional payments which, after an appropriate period, oblige farmers to continue to use environmentally-favourable practices without ongoing financial assistance. It should be noted, however, that this recommendation does not exclude the payment of income support payments, nor subsidies for practices which enhance landscape amenity and other environmental values, nor does it exclude the orientation of other agricultural policies towards the simultaneous production of environmental benefits.

Regulatory, as well as economic approaches can be used to achieve compliance with the Polluter-Pays Principle. Regulations impose most pollution prevention and control costs on farmers and have the advantage that they can be more precisely targeted than economic instruments. Enforcement of regulations, although unpopular, is essential for compliance with the Polluter-Pays Principle and needs to be strengthened in most OECD countries. In addition to a number of policy changes in some cases this may require greater budget allocations for enforcement and, in other cases, a change in administrative practice.

Products

Product regulations, standards and marketing arrangements to protect human health exist in all OECD countries and indirectly they have a significant impact on the quantity of pesticides used during production. Integrated pest management may be possible if the grading standard for premium grade fruit is set at one level, but may be impossible if it is set too high. Consequently, during the review of marketing and grading standards full account should always be given to the likely impact of product marketing standards on pesticide use.

Recently, there has been growing demand in several countries for organically grown produce or health food. Although the development of this market is still in its infancy some people are prepared to pay a significant premium for such food. Organic farming is usually compatible with protection of the environment and with human health objectives. Standards for the sale of organic food, however, vary from country to country and opportunities exist for the further development of this market which, if it becomes significant, can be expected to lead to some reduction in agricultural pollution. Consequently, there is an emerging need to develop a set of international guidelines for the production, marketing and labelling of organically grown products so that no country obtains an unfair trading advantage and organic farming can be further promoted. Higher education, research and advisory programmes should also be adapted to make extension officers more aware of this new development and to inform farmers of opportunities to diversify into organic food production.

Capital structure and development

As indicated at the start of this section, agriculture is going through a period of rapid structural change and development. Apart from the instruments and approaches mentioned above, taxation and regional planning policies are increasingly being used to maintain and improve landscape amenity values in rural areas. They have the advantage that they also require consideration to be given to other regional development objectives and are particularly effective in protecting rural areas from adverse development.

Many distortions exist within tax policies in most OECD countries. Examples include land tax exemptions for agricultural but not forestry land and the provision of income tax concessions for soil conservation and the establishment of pine plantations. Some of these distortions benefit the environment but others have the opposite effect. There appear to be major efforts in many countries to remove such environmentally negative incentives from tax policy. Other work in the OECD, however, suggests that taxation is a policy instrument which should be principally reserved for collecting revenue and income redistribution goals and that consequently income, capital and land tax policies should be neutral to the environment and agriculture. Inheritance and property transfer laws can also have negative effects on the environment and tend to hinder structural adjustments which can improve environmental quality. Although of short term advantage, in the long run opportunities for economic growth, the sustainable use of agricultural resources and environment quality are likely to be greater if tax incentives and concessions are not used as policy instruments to directly influence environmental and agricultural objectives.

Research and education

Opportunities also exist for the strengthening of research and advisory programmes so that greater emphasis and broader consideration is given to environmental objectives. Consideration should also be given to rearranging some research away from traditional agricultural products and practices towards those which enhance agriculture's contribution to environmental quality.

To this end, countries have begun to appoint environmental representatives to assist with the allocation of research funds and, wherever possible, to ensure that the environmental as well as agricultural implications of technological advances are evaluated by researchers.

As awareness of the environmental consequences of different agricultural practices is a necessary prerequisite to the development and implementation of policies which give full consideration to the effects of agricultural policies on the environment, opportunities also exist for expanding the environmental content of university and other training courses for agriculturalists. Similarly, as environmental administrators are becoming more involved in agricultural policy formation, their professional training should be adapted to contain more information on agriculture.

With regard to the effects of environmental pollution on agriculture and agriculture on the environment, it is apparent that there is a need for further research on:

-- the impact of air pollution on agricultural production and particularly the likely benefits to agriculture of reducing these impacts;

-- integrated pest management and other environmentally-favourable agricultural practices;

-- the likely impact of new technologies on the environment;

-- crops that use less environmentally damaging inputs;

-- developing plant varieties which use less environmentally damaging inputs;

-- practices that cause the accumulation of pesticide residues in groundwater; and

-- the avoidance of soil pollution from heavy metals.

Such calls for further research, however, should not be regarded as an excuse for inaction. In almost all cases, there is now sufficient information available for countries to begin to address the urgent problems which face agriculture and the environment. For example, while there may be a need for further research on the effects of nitrate pollution of water supplies and the accumulation of heavy metals in soils, it is clear that these effects are negative, and many governments have already taken action to reduce this pollution. In most of these cases prevention may well be easier and less costly than cure.

Monitoring

Systematic monitoring, the subsequent objective assessment of the data collected and the timely diffusion of the resultant information are necessary preconditions for successful integration. Without such information, it is often difficult if not impossible to adequately assess the likely effects on the environment of a policy. Information from countries has revealed that there is a particular need to improve the quality, relevance and availability of information needed for decision-making. There is also a need to further develop the collection of information on trends in the state of the environment in rural areas.

6. EMERGING OPPORTUNITIES FOR THE SUCCESSFUL INTEGRATION
OF AGRICULTURAL AND ENVIRONMENTAL POLICIES

Environmental considerations provide added impetus to the reform of agricultural policies. Opportunities exist and are emerging for the integration of agricultural and environmental policies during the reform process. The purpose of this final section is to point out the wide range of opportunities for integration which could stem, in particular, from the agricultural policy developments currently under discussion. Some of these reforms will take time to come about, but many could be realised within the next few years. All of the opportunities identified in this section are already either being implemented or under serious discussion in one or more OECD countries.

Within the context of this section it is suggested that environmental policy-makers should carefully monitor proposed changes to agricultural policy and ensure that any substantial environmental ramifications and opportunities are given full consideration. Likewise the onus is on agricultural policy makers to take environmental considerations into account at an early stage in the development of policy reforms for agriculture.

OECD countries are at a crucial stage in reforming agricultural policies and, in particular, are seeking ways to improve trade in agricultural products and overcome the problems associated with agricultural support policies. It is expected that anticipated policy reforms could substantially alter the location and intensity of agricultural production, thereby significantly affecting the quality of some rural environments. But at the same time it is recognised that these reforms offer a number of major new opportunities simultaneously to achieve agricultural and environmental objectives in a manner which is mutually beneficial to farmers, the environment and society in general.

Many countries are already moving towards the reduction of protectionist support measures and the revision of existing production incentive programmes. Approaches being evaluated include: the reduction of price support and other production incentives; the reduction of existing quotas on production; the provision of income support to farmers; the use of environmental payments; support for diversification into alternative crops; the use of set-aside programmes; the introduction of quotas for crop and livestock production; and imposing levies, taxes and quotas on inputs. In each of these cases it is clear that many opportunities exist to orient these changes so that they both achieve agricultural objectives and enhance environmental quality.

From an environmental point of view, the issues to be considered are the extent to which each of these measures will a) improve nature conservation and landscape amenity values; and b) reduce agricultural pollution by decreasing production intensity at a farm and regional level. This latter effect, which is known as "extensification", is seen by many to offer the greatest chance of improving the environment, increasing economic efficiency and reducing agricultural surpluses.

Production incentives

Price support and international trade and tariff arrangements have had a dominant influence on the location, intensity and size of most agricultural industries. Tariff barriers have been used, for example, in many countries to maintain internal prices above international prices and thereby maintain and in some cases substantially increase agricultural production. One disadvantage of high support prices is that they provide the greatest benefits to the largest and most efficient producers who are often responsible for agricultural pollution and the least to those in marginal and less favoured areas where support often brings greater environmental benefits. It is also true that high price support policies in some instances have encouraged cropping, for example, in some marginal areas generally unsuitable to continuous cereal cropping and thereby increased the risk of soil erosion and environmental damage. Nevertheless, although other more efficient support mechanisms exist, it is true that in marginal areas this approach of artificially raising internal prices has had the benefit of preserving an agricultural landscape in areas that would otherwise have been left in their native state or under forestry.

Differential prices and other incentives have also been successfully used to maintain production in areas where agriculture is considered essential for regional development, landscape quality and for defence reasons. From this, it is apparent that unless replaced by compensating measures, a reduction of support measures in such less favoured and environmentally -sensitive regions could lead to significant environmental losses.

It has been agreed by the Governments of OECD Member countries at Ministerial level that reductions in agricultural support, together with other appropriate measures to allow market signals to influence the orientation of agricultural production, need to be implemented. Some countries, such as New Zealand, have significantly reduced the extent of production incentives provided to agriculture, but the reform process throughout OECD countries will take time.

Generally, a reduction in production support measures can be expected to lead to general environmental improvement through the reduced use of inputs such as pesticides and fertilizers and, also, the induced structural changes which are likely to lead to the adoption of less intensive production practices. A reduction of support may have very different effects, however, depending upon whether or not environmental policies are simultaneously implemented in all areas. This environmental improvement will be strongest in intensive production areas, which under the revised price regime can be expected to lower production intensity, but to continue to produce the same or similar crops. Price reductions which cause changes in crop rotation patterns or the introduction of completely new crops might also be expected to result in less agricultural pollution, but if they induce substantial structural change could have an adverse effect on some landscapes.

In the extensive production areas and environmentally sensitive regions, however, environmental gains would result from the reduced use of fertilizers and pesticides, but extensive structural change can be expected to occur. Consequently, in these areas measures to protect landscape and nature conservation values may be needed. These latter regions, however, account for a relatively small proportion of agricultural production.

A fall in price stability associated with a reduction in price support is also likely to bring environmental improvement since farmers have, traditionally, attempted to protect themselves against the risk of price falls by diversifying their production. The more the risk of price variability is reduced by price guarantees, the greater the incentives for specialisation and the intensive use of inputs. Environmental gains can be also expected from reduced production incentives for cropping on highly erodible areas and wetlands. In summary, if a reduction in price support is to bring about net social gains in all areas then there will also be a need for careful attention to be given to landscape amenity, nature conservation and community development objectives in areas of special environmental importance. This, however, is not an argument for not realising the benefits from reducing price support.

Cross-compliance

If a decision is made to continue with price support and other production incentives on either a temporary or permanent basis, one attractive and recommended approach is to require that recipients adopt and comply with a set of prespecified land-use conditions and practices in return for this support. This is known as cross-compliance.

Its implementation, however, would require a transition from indirect market support to arrangements which permit the selective payment of support to individual farmers. Deficiency payments, for example, can easily be linked to cross-compliance provisions while indirect market support cannot be linked to a requirement that farmers use environmentally-favourable practices. Consequently, a general move towards the introduction of cross-compliance -linked deficiency payments is seen as an environmentally-desirable direction for agricultural policy to take in reducing support to agriculture.

Similarly, income support could be linked to cross-compliance criteria. Both these approaches have the additional advantages that unlike indirect market support arrangements, they can be targeted to areas where social and environmental needs are greatest.

In opting recently to continue to offer deficiency payments to farmers, the United States has decided to exclude from continuing financial assistance programmes any farmers who expand production into wetlands or onto highly erodible soils without an approved conservation plan. These exclusions are known as the "sodbuster" and "swampbuster" provisions. They establish the general principle that all production-inducing measures should be framed so as not to induce a deterioration of environmental quality.

Income support

One alternative to the continuation of production-incentive policies, such as price support and selective tariff arrangements, is the introduction of direct income support. The OECD agricultural trade study (1987) urged that domestic policies be reformed to allow for a reduction in support granted to the sector through output-related measures (Press Release PRESS/A(87)38).

This approach involves some price reduction with a resultant reduction in the production intensity of agriculture. Environmental benefits could be expected in areas where, for example, there is ground water pollution from high fertilizer use. Above all, it would also make it less profitable to continue with existing agricultural practices in marginal areas.

As indicated above, the reduction of production incentives without the introduction of incentives to protect the environment, can have substantial effects on landscape quality and regional development objectives. One way of doing this is to introduce income support schemes which are conditional upon the maintenance of environmental benefits. Without this clear coupling of income support with environmental protection and improvement the outcome of income support policies will largely depend on how and where farmers chose to invest the income support they receive. Consequently, it would seem more attractive to target income support towards regional development, and the environment. This approach has the additional advantage that it does not involve the direct, and hence more visible, payment of social security to farmers in preference to other perhaps more needy groups. It can also be used to encourage farmers to diversify into other non-agricultural activities such as the provision of accommodation for tourists.

Coupling of income support to the environment and other related activities offers gains from the general extensification of agriculture at a farm and regional level, and also an opportunity to maintain environmentally -desirable agricultural practices and at the same time achieve social welfare and other objectives associated with the maintenance of farm incomes.

One option would be to offer farmers the opportunity to become involved in various programmes directed to environmental maintenance. The use of management and other similar agreements to encourage farmers to maintain desirable but otherwise unprofitable practices, which without an agreement would not be profitable, is a recommended approach. The emphasis, however, is on reimbursement for additional maintenance costs and expenditure required under the agreement. Moreover, such agreements should always have a positive orientation and should be used to pay farmers to adopt practices which enhance the environment as opposed to those aimed at pollution reduction.

An alternative approach, which is attractive for its administrative simplicity, is to provide income support in proportion to the area of land held by each farmer. In the long run, however, the environmental benefits of such an approach are likely to be few as its main effect will be to retain people on their farms, but not to influence the management practices they adopt.

Support for diversification

In recent decades, agriculture has tended to become highly specialised and in many regions monocultures have become the norm. This has been due, in part, to price stability and, in part, to the nature of the production incentives offered to farmers. If this situation is to change then a difficult choice has to be made between maintaining these incentives or the provision of others to encourage greater production diversity.

Greater diversity in production can be expected to improve landscape amenity and nature conservation values and, with appropriate controls, reduce agricultural pollution. Opportunities for enhancing diversification include the provision of subsidies for the establishment of hardwood forests, payments to assist farmers during the transition to organic farming, payments in return for a substantial reduction in the use of inputs and grants to assist with the development of alternative crops or which aid farmers to become more actively involved in the provision of tourist and recreation facilities. It is recognised, however, that there are dangers in always providing subsidies to farmers.

Caution is offered against the introduction of further production incentives. In the long run, it is likely that the removal of the policy distortions which in the past have prevented diversification would be a more efficient approach. Generally, from an environmental viewpoint schemes which encourage the diversification of agriculture can be expected to be environmentally favourable. Experience in other sectors, however, suggests that these schemes will be most effective if assistance is provided only for the difficult transition phase and a clear time limit is outlined at the commencement of the programme.

Set-aside policies

An alternative method of reducing agricultural production, which has been used in a number of countries for many years, is the implementation of land set-aside schemes. There are many versions of these schemes and hence caution must be used in discussing them. Some work on an annual basis, others for between 10 and 50 years. Some are targeted specifically to produce environmental benefits. Drawing on the information described in this report, it would seem that set-aside programmes, which reduce the use of inputs and environmentally damaging practices in sensitive ground water recharge areas or areas of high ecological value, offer an invaluable opportunity to overcome a number of short-term agricultural production problems and enhance environmental quality. It is also noted that the expectation of net environmental benefits from set-aside programmes will be enhanced if these programmes are coupled with reductions in existing price support arrangements. Initially most set-aside policies were targeted to marginal production areas but increasingly it is being realised that budgetary savings will be greater, slippage problems less and environmental benefits greater if set-asides operate in key production areas. One example of this is the recent proposal to set-aside filter strips along stream banks in the United States. The benefits of such schemes, however, are limited to the areas which are actually set aside from production or diverted to another use.

To be effective these set-aside programmes will have to be implemented, however, in a manner that does not lead to soil erosion or decrease water quality. In some areas, incentives may also need to be provided for the adoption and retention of environmentally-favourable agricultural practices without which farmers may have no reason to continue to care for these lands. In most schemes, it is normal practice however that the payment for set-aside is conditional upon the continued maintenance and care of the set-aside land. In the United States experience is revealing that it is much easier to achieve care and maintenance by calling the set-aside land a "conservation reserve".

This title makes it clear that even though the land is set-aside it is the duty of the farmer to care for this land in a manner which enhances it conservation value. As a result there is considerable community pressure for farmers to manage them in a way which enhances landscape and nature conservation values. A related conclusion is that there is greater flexibility and efficiency of selecting land via a tender system with payments in proportion to anticipated environmental benefits and budgetary savings.

An interesting European Community development is the introduction of extensification schemes and a set-aside programme. Under these schemes it will be possible for farmers to receive payments in return for agreeing to reduce production by at least 20 per cent for 5 years. If cereal land is taken out of production, it must be fallowed, afforested or put into non-agricultural use. Experience in other northern European countries and in the United States has revealed that bare fallowing can increase nitrate leaching and soil erosion problems and can cause problems with respect to landscape amenity and conservation objectives. The experience with rotational fallows where the actual area which is set aside from year to year changes have similar environmental problems and, as well as this, serious administrative impediments. The experience also suggests that if the period was extended to apply for a period of say 10 years there would be a much greater incentive for farmers to diversify into and develop alternative enterprises. When implemented throughout the European Community, it will be important to ensure that these national schemes contain provisions which ensure that they do not increase nitrate leaching and also do not unintentionally reduce landscape amenity, nor make it more difficult to achieve conservation objectives.

Input and output quotas

An alternative and potentially complementary approach to the reduction of production incentives is to introduce limits on production. Such limits take many forms and, in the past, have often encountered many problems in implementation.

Production quotas, for example, are often used in the dairy industry. Quotas are also used to limit the area of land which may be planted to a crop and the number of animals which may be run on a farm. Production quotas have both merits and drawbacks. Generally they limit the further development of any negative effects which agriculture has on the environment but, at the same time, they tend to encourage a greater intensity of and specialisation in production on those farms which remain in the industry. Although further environmental deterioration is prevented, few environmental gains can be expected from the introduction of quotas.

The other way of influencing production is to restrict the use of certain inputs. In the case of scarce water resources, providing the use rights are transferable, input restrictions have been a very successful approach to improving water quality and the environment in general. When input quotas are transferable and acquire scarcity value, however, they can facilitate more speedy adjustment and promote diversification.

Experience with other agricultural inputs is almost nonexistent. Where an input such as nitrogen fertilizer is a principal source of pollution, the introduction of nitrate quotas as a means of reducing agricultural surpluses should not be discounted and in areas where pollution is a problem they should be seriously evaluated because of their favourable impact on farm income. The net environmental benefit of input quotas, however, is likely to vary from region to region and depend heavily on the changes in crop rotation and other practices which they induce.

* * *

REFERENCES

OECD (1986)   OECD and the Environment, OECD, Paris, 220p.

OECD (1987a)   Pricing of Water Services, OECD, Paris, 145p.

OECD (1987b)   National policies and agricultural trade, (OECD PRESS/A(87)38).

OECD (1988)   Country approaches to the integration of agricultural and environmental policies, OECD, Paris, (in Press).

World Commission on Environment and Development (1987) Our common future. Oxford University Press, Oxford.

European Community (1985a) Perspectives for the Common Agricultural Policy: The Green Paper of the Commission.

European Community (1985b) A future for Community Agriculture: Commission Guidelines.

ANNEX 1

COUNTRY APPROACHES TO THE INTEGRATION OF AGRICULTURAL
AND ENVIRONMENTAL POLICIES

1.   INTRODUCTION

Overview

Agriculture, one of the major economic sectors of OECD countries, has undergone a revolution in the 40 years since World War II. The technological and scientific explosion of the past four decades has literally transformed agriculture in the industrialised countries. The resultant increase in agricultural productivity has also created and released resources which, in turn, have substantially stimulated economic growth in other sectors.

New plant strains and animal breeds, new pesticides and fertilizers, the introduction of mechanized farming, large-scale irrigation, new marketing arrangements, and above all, vastly improved knowledge and management skills have all contributed to nearly doubling overall agricultural production in OECD countries in the past 35 years. Countries are now generally self sufficient in the production of basic food products and a much greater diversity of high quality food products exists.

Perhaps the most visible impact of agricultural modernisation has been a socio-economic one, in terms of changed rural lifestyles, notably the decline of the number of family farms, and the progressive depopulation of agricultural areas. Many countries have undergone fundamental demographic and economic changes as capital inputs have increasingly displaced man's labour, making the management of increasingly large tracts of land feasible and economically efficient. Farming has been transformed from a largely labour orientated production system to one orientated towards the use of technology. Rural depopulation, the scale of farming, its mechanisation and its intensification, are now leading to problems in some areas for wildlife preservation and landscape management.

There have been other, less visible changes such as groundwater pollution and the accumulation of heavy metals in soils which have resulted in costs to the natural and human environment.

There is remarkable consistency in the kinds of environmental problems associated with agriculture that have been inventoried in many OECD countries, although their wide geographical and climatic diversity means that the relative importance of different environmental problems associated with agriculture varies widely. Animal manure in semi-arid Australia is not the problem that it is in sub-Arctic Finland. Problems associated with the density and proximity of urban and agricultural lands in Japan and the United Kingdom are far less criticial in sparsely populated Canada and New Zealand. Problems of agrochemicals are more acute in European countries with high population densities practising widespread intensive agriculture than they are in countries like Canada, the United States and Australia where different climatic and market conditions make more extensive practices economical.

The trend towards increasingly large-scale agriculture has, in some cases, transformed woodlands, mountain slopes and wetlands. The extensive use of agricultural chemicals, in the form of fertilizers and pesticides, while dramatically increasing yields, has meant that these chemicals are also increasingly present in the environment, in water, in food, in soil and in the air, long after their immediate agricultural purpose has been served. Problems such as soil erosion, the nitrate pollution of ground and surface waters, the eutrophication of lakes, estuaries and coastal waters, and the pollution of soils by heavy metals are becoming more acute. Subjected to increasingly intensive farming techniques, some lands can no longer absorb and recycle the inputs, byproducts and waste involved in agricultural activity.

Agricultural research developed a plethora of chemical products to control plant, animal and insect pests. Aided by modern analytical techniques the problems associated with pesticide residues in feed for animals and in food for human consumption have become more apparent. Traces of the insecticide DDT, detected in the high arctic, provided a first dramatic example of pesticide distribution and accumulation in the environment on a global scale.

Livestock raising has undergone a similar qualitative and quantitative evolution, as the intensive husbandry of pigs, poultry and veal replaced pasture and fodder raised animals. The output potential of livestock has been greatly enhanced through the use of artificial feed additives, veterinary medicines, genetic improvements, biotechnology and growth-stimulating hormones. But it has created problems in managing organic waste storage and treatment, leading to pollution of surface and groundwater. The intensification of agriculture and specialisation in outputs are among the principal traits of modern agriculture and among the principal sources of related environmental problems.

Fundamental perceptions about agricultural activity and about the farmer, as beneficiary and protector of the natural environment, as an economic actor living in symbiotic harmony with the land in an increasingly urbanised world, are beginning to alter as the significant, long-term environmental impact of agricultural activities on the landscape, on drinking water, the air we breathe, and the food we eat is increasingly recognised.

Most governments now accept that there is inevitably some trade-off between environmental quality and otherwise essential and beneficial agricultural activity. But questions are being raised about the logic of sustained pressure on the environment to produce agricultural surpluses. This is now being seen as one of the most paradoxical phemonena of agriculture in OECD countries, and a question worthy of extensive debate.

The recognition of the positive environmental role of agriculture in recycling nutrients, maintaining wildlife habitats, etc. justifies the growing concern about the extent to which agriculture is being seriously affected by the impact of environmental phenomena arising from other sources. Entire regions of Europe and North America are being subjected to the effects of acid deposition. Prime agricultural land is being submerged under urban sprawl. The tremendous implications for agriculture of global climatic change due to ozone depletion and carbon dioxide concentration are only now beginning to be understood. The impact of the Chernobyl accident also has had a significant impact on agricultural production in some OECD countries, and above all has heightened awareness of the threat to agriculture from large-scale releases of radioactive materials.

Work is underway in various countries to assess the cost of damage to agriculture of pollution from other sources, as well as the impact of pollution arising from agricultural activity and resource degradation. At this stage little information is available. It is estimated, however, that to keep all drinking water in the United Kingdom within the European Community limit of 50 mg of nitrate per litre would require the expenditure of £200 million in capital costs and an annual expenditure of £10 million in 20 years time. To keep all water within the current government limit of 80 mg per litre would require £40 million of capital expenditure and running costs of £3 million per annum. In Canada, the direct on-site cost of resource degradation on farms is estimated at C$1 billion annually. In the United States it has been estimated that the downstream cost of soil erosion from American farms is probably in the vicinity of US$2-6 billion per annum. No estimates of the positive value of agricultural practices and the contribution they make to the environment are reported in the country information papers. The impact of pollution on agriculture will be considered in detail on the basis of an OECD-wide review in the second part of this project.

"Third generation" environmental policies

Environmental policies in OECD countries have evolved considerably in the past two decades. The first generation began in the sixties and was conceived in terms of reactive damage-control and clean-up programmes. The second generation gradually evolved in the seventies towards a more preventive, anticipatory mode. As the most urgent problems were resolved, a third generation is emerging as policy-makers turn their attention increasingly to developing mechanisms which ensure that environmental factors, positive and negative, are an integral part of policy formulation, development and implementation.

This so-called "third generation" environmental policy has begun to receive increasing attention both at the national and international level.

Integration of agricultural and environmental policies -- OECD's role

Thus 16 years after the OECD began work on problems of environmental policy in its Member countries, governments, at a meeting of Ministers of Environment in June 1985, recognised the need to proceed towards a systematic integration of environmental policies with other polices, and declared "that they will: Ensure that environmental considerations are taken fully into account at an early stage in the development and implementation of economic and other policies in such areas as agriculture, industry, energy and transport."

At the same time, Ministers agreed that the OECD should undertake work in this area and, subsequently, governments approved an initial study on the agricultural sector, in part because of its overwhelming role in shaping the landscape and the natural environment as a whole, in part because of growing concern with the increasingly apparent serious environmental impacts of modernised, intensive farming and also because of the recognised environmental benefits associated with many agricultural practices.

This draft report, a synthesis of country information papers and the first result of OECD's work on the Project "Integration of Environmental Policies with Agricultural Policies" describes the current approaches in OECD Member countries to the problems of integrating agricultural and environmental policies. Ongoing work will explore specific policy areas; fill information gaps; identify additional successful approaches to the more effective policy integration; and identify the fundamental institutional, societal and economic conditions which lead to the better integration of agricultural and environmental policies.

Approach and methodology

Using guidelines developed by the ad hoc Group on Agriculture and Environment, Member countries were invited to submit information on the approaches they are taking and the experiences they have gained in integrating environmental policies with agricultural policies. Each submission briefly situates agriculture in a national context; describes existing and planned institutions and policy mechanisms for integration; and then outlines the extent of policy integration with respect to specific aspects of agricultural policy such as inputs, practices, products, and capital structures, as well as the response to known positive and negative environmental effects of agriculture. In the light of the significant impact, particularly in northern Europe, of the Chernobyl nuclear reactor accident, participating governments were also invited to comment on implications for agricultural policy of the risk of nuclear accidents.

This report, drawing upon the country information papers, summarises the approaches which countries are taking to integrate agricultural and environmental policies. In the interest of brevity, the Secretariat has endeavoured throughout the text to use the most appropriate examples of the approaches which countries are taking in preference to providing an exhaustive list. Subsequent work by the ad hoc Group will examine specific policy areas and evaluate the effectiveness of the different approaches being taken by Member countries.

2. POLICY FORMATION, IMPLEMENTATION AND REVIEW

General context

It is apparent throughout the country information papers that a number of fundamental considerations underlie policy formulation, implementation and review. The first of these considerations is the initial inherent dichotomy between long established agricultural institutions and their newer environmental counterparts. The second is the need to take into account the myriad actors involved in the agricultural and environmental sectors.

The third feature of policy in this area has been, as noted above, the relatively recent identification and appreciation of many of the problems to be addressed. Consequently, urgency has often been attached to finding policy responses to critical issues. The fourth is the need to balance the environmental and overall economic cost of present agricultural policies against their political, social and private economic benefits.

Since World War II, agricultural policies throughout the OECD area have almost invariably aimed to increase productivity and production (generally in terms of self-sufficiency or export goals), while seeking to maintain and enhance the income and social conditions of the farming population. But although the objectives have been uniform the approaches taken and the policy instruments used to achieve them have varied widely. Only recently have these policy objectives begun to change on economic grounds, in the face of growing surpluses and the increasing costs through high farm prices or direct subsidies of maintaining farm incomes. The positive and negative impacts of these approaches and instruments on the environment are also being recognised and the causes of these impacts debated.

Over the past two decades, an entirely new set of policy considerations related to the environment has had to be incorporated in, reconciled with, and in some cases, imposed upon, long-standing agricultural policies and institutions. Different kinds and degrees of response have been developed to the need to reconcile and integrate agricultural and environmental policy objectives. These range from highly sophisticated administrative structures founded on an early recognition of the problems and a clear committment to action on the part of government, through less well co-ordinated, piecemeal solutions in states where the severity of the problems have been less marked.

In the 1970s protection of the environment was rarely an explicit objective of agricultural policies, some early exceptions being Norway and Sweden both of which revised their Agriculture Policy objectives in 1977 to include environmental considerations. Similarly, since 1968 all Ministers, government departments and public bodies in the United Kingdom have had statutory conservation duties. Stimulated by the National Environmental Protection Act (NEPA) which was enacted in 1970, the United States also began to define environmental protection as an agricultural objective during this period. One of the more interesting observations in the country information papers is that environmental objectives are in fact being served by policies designed to limit agricultural production in certain areas because they have the effect of lessening negative environmental pressure on the land.

As noted above, another fundamental feature of policy making in the environmental and agricultural spheres in OECD countries is the broad range of actors involved. It is not surprising that in a sector that is so important to national economies, there should be plethora of agencies at national, regional and local levels (not to mention supra-national organisations like the European Community) who must inter-act with farmers, their associations and representatives, the producers and suppliers of farm inputs and the buyers of farm products, consumer groups, the media and the scientific community.

Moreover, there are private groups, including non-profit foundations such as those found in Germany, the United States and the United Kingdom, special interest environmental groups, and indeed full-fledged ecology-oriented political parties, all of whom have some capacity to influence policy. One interesting measure of the difference in approach to policy making among OECD governments is the extent to which these non-governmental organisations are invited to participate in policy formulation and development. Such participation is actively encouraged in Canada and the United States, for example, in the context of formal environmental assessment review procedures which call for formal comments from the public during the decision making process.

Institutional frameworks for agricultural and environmental policy formation

Pressure to modify agricultural policy for environmental reasons is, as we have seen, a relatively recent phenomenon, the impact of which is only beginning to be felt. As the level of awareness of the sector-wide implications of agriculture and environment problems has increased, formal voice has been given to the issue in the form of clear policy statements by governments, implying that in many countries, a certain degree of consensus about the need to integrate agricultural and environmental policies has been reached. Examples of these include New Zealand's "Position statement on Environment"; Sweden's recent Environmental Protection Board policy statement on agriculture; The Netherlands' indicative multi-year programme for environmental management; and the United States Department of Agriculture's Environmental Policy Statements on issues such as fish and wildlife.

Over the past two decades, most OECD countries have created agencies or ministries specifically charged with the development and implementation of policies and programmes to monitor and protect the environment. Some of these are of very recent date: Finland (1983), Sweden (1986), Germany (1986), Italy (1986), and New Zealand (1987). On the other hand, the historical place of agriculture has meant that ministries and departments of agriculture have been in existence for decades in all OECD Member states.

Within government both at the national or regional level, a variety of departments, regulatory agencies, advisory groups and boards are involved in agricultural policy and correspondingly have a role to play in developing and implementing related environmental policy. Typically in OECD countries, these include, in addition to agricultural departments, ministries and agencies responsible for water, forestry, energy, natural resources, public works, tourism and economic planning. These structures often are mirrored at the regional level.

A marked feature of approaches by OECD countries to policy making in this area has been the extent to which new, innovative structures and arrangements have been developed within a relatively short time frame. The Country Information Papers describe a broad cross-section of such mechanisms designed to improve the integration of agricultural and environmental policy considerations. Examples include:

-- parliamentary committees appointed to develop policy options and proposals which simultaneously achieve agricultural and environmental objectives;

-- problem specific co-ordinating committees, task forces and working groups, often focussed on issues like soil, forests, land, water, acid deposition and toxic chemicals featuring broad governmental representation, and occasionally including the private sector;

-- national policy councils and advisory bodies that bring in all concerned branches of government at national and regional levels as well as the private sector to determine guidelines for the implementation of policy in specific sectors like nature and soil conservation;

-- mandatory procedures for inter-agency consultation in the development of new policy and legislation;

-- formal inter-agency agreements committing signatories to collaboration in policy implementation;

-- formal agreements between national and regional government agencies aimed at specific problems;

-- agreed "frameworks of objectives" to facilitate co-ordination and complementarity of policy making at national and regional levels;

-- clearly identified environmental units established within agricultural ministries;

-- the extensive delegation and devolution of environmental responsibilities to regional and occasionally local governments;

-- municipal environment and agriculture committees mandated to develop "bottom up" policy proposals for possible inclusion in regional and national plans;

-- quasi-governmental bodies such as "countryside commissions", nature conservation councils and trusts with statutory rights to intervene in the policy making process; and

-- transfers and exchanges of expert personnel between agricultural and environmental agencies.

The required integrating and co-ordinating mechanisms can thus be said to be horizontal across legislative and executive bodies at the national level. Both vertical and horizontal coordinating mechanisms are needed through regional and local administrations and, ultimately, to the farmers themselves, as well as the rest of the actors in the agricultural Community.

Integration and co-ordination at the national level

Austria, Canada, The Netherlands, the United States, the United Kingdom and Switzerland provide examples of a highly evolved institutional response to the problems at the agriculture and environmental interface. Well established national environmental ministries or agencies have had the time to develop their structures and attain a position of relative influence within their administrations vis-a-vis the historically important and powerful economic and resource departments, including agricultural ministries. In all these cases, formal interdepartmental machinery in the form of committees, task forces, working groups, or issue-specific inter-agency groups have been created, such as the Canadian Federal government's inter-departmental committees on Water, Land, and Toxic Chemicals. Other horizontal mechanisms include the formation of advisory councils which are often used to help develop research priorities in countries like Australia. Indeed, such inter-departmental co-ordinating mechanisms are common to virtually all OECD governments, to a greater or lesser extent. Perhaps Denmark's Committee for Farming Policy is the most explicitly integrated mechanism, grouping the Agriculture and Environment ministries as well as agricultural organisations, research institutes and the

Society for Nature Conservation. Another example is the Austrian committee for soil protection which contains representatives from all the relevant ministries, the laender and the universities.

The Netherlands, since 1984, has aimed towards the development of a cohesive environmental policy through the annual preparation of an indicative five-year plan of action negotiated by three different departments and issued under the joint signatures of the Minister of Housing, Physical Planning and Environment; the Minister of Agriculture and Fisheries; and the Minister of Transport and Water Management. In preparing these plans these Ministries have developed a two-track "source-effect orientated" approach. Agriculture is one of the target groups for source orientated policy and measures being taken to discourage practices which cause problems such as nitrate pollution.

Another mechanism of collaboration between agricultural and environmental agencies at the national level takes the form of formal agreements, such as the Memorandum of Understanding between the United States Department of Agriculture and the Environmental Protection Agency.

The relative influence of environmental agencies via-a-vis their agricultural counterparts is perhaps illustrated by the extent to which regulatory control for certain aspects of agricultural activity has passed to the former. The Environment Protection Agency in the United States was one of the early environment agencies to be given direct responsibility for pesticide registration. In many other countries, however, responsibility remains with the agricultural authorities. There does appear to be an underlying trend in OECD Member countries, notably in Finland, Italy and France, whereby once environmental ministries have been established, many policy responsibilities are progressively transferred to them from existing agricultural agencies.

Japan has a Ministry of Agriculture, Forestry and Fisheries and an Environmental Agency. The principle responsibilities of the Ministry are to ensure a stable food supply and rural development while the Agency is responsible for environmental protection and the prevention of pollution. Amongst other things, the Ministry of Agriculture, Forestry and Fisheries is responsible for the registration of pesticides.

Sweden, Switzerland, The Netherlands and Denmark have opted for highly formalized approaches to the integration of agricultural and environmental policies by enacting mandatory consultation procedures, which require agricultural and environmental agencies to collaborate in the development of new policies and legislation.

In Turkey, where a national environment ministry has not yet been established, a General Directorate for Rural Services has been created within the Agriculture Ministry with a broad, cross sectoral mandate to deal with all factors affecting agricultural productivity and the protection of the agricultural resource base.

France and the United States cite the important role that can be played by exchanges of personnel or by transferring officials formerly associated with agricultural activities into newly created environmental agencies. Temporary transfers enable the concerned agencies to develop a body of personnel who understand each other's objectives, policies and operations.

Studies and assessments which examine the future of agriculture and agriculture's impact on the environment are helpful integrating devices. Under the Resources Conservation Act the United States studies trends in agriculture and resource protection and adjusts its programmes to solve present and emerging problems. As a result of a recent study, priority was placed on reducing soil erosion and, at the same time, improving water quality.

Integration and co-ordination among national and regional administrations

The decentralisation of policy making responsibility from central to provincial, state, county, laender and other sub-national levels of government is a phenomenon common in one form or another to all OECD governments. Although perhaps most institutionalised in so called Federal States, the role of regional authorities can be just as important in French "départements", Danish counties or Finnish provinces. The decentralisation of policy making from central to regional levels of government is reflected at the level of the European Community, where, as part of its response to the Single European Act, the Commission has stated its wish to replace the concept of 'projects' with that of multi-annual 'programmes', within which Member States would decide and implement their own projects. Nonetheless, a fundamental rule of thumb throughout the OECD area that agricultural and environmental policies deemed to be national in scope are the prerogative of central governments. Pesticide regulation falls clearly in this category. In several countries, notably France and Greece, so-called regional administrations are in fact decentralised arms of central government departments, representing a functional as opposed to a political or geographic decentralisation. Decentralised and dispersed responsibilities for agriculture and the environment mean that co-ordinating mechanisms are required not only at the national level among departments and agencies, but also between federal and regional administrations.

Vertical integrating mechanisms among national and regional governments take a variety of forms. In Australia's federal structure, all tiers of government (including local governments) coordinate their actions via sector-specific Councils for agriculture, environment, nature conservation, water resources, and soil conservation.

As between federal agencies there exist many examples of formal agreements between national and regional levels of government respecting agricultural and environmental concerns. Examples of vertical integration in the United States include:

-- a non-point source task force and a ground water advisory group which brings together all interested agencies and is composed of officials at all levels of government and the private sector; and

-- the Chesapeake Bay programme which is guided by a management group comprising federal and state Officials and with strong participation from private sector representatives, including farmers.

Among the countries with a high degree of institutional recognition of environmental and agricultural problems, Austria, Canada, the United States Switzerland and Australia also accommodate federal structures, whereby responsibility for policy making in both the agricultural and environmental fields is constitutionally shared with or devolved to state and provincial authorities. In all these countries, many national agricultural and environmental policies are implemented directly by states or provinces. In each case administrative arrangements have been found to ensure that their views on priority, feasibility, and workability are reflected in national policies and programmes.

Thus, in the United States for example, enforcement of pesticide regulations has been devolved to "qualified states". These qualified states have the authority to set standards more stringent than those adopted at the federal level, not only for pesticides but also in regulating air and water pollution. Similar provisions exist in Canada. In other countries such as Germany, it is also common for the federal government to pass general framework legislation and then leave regional governments to further define and implement laws and regulations in a manner which takes account of regional differences.

Land-use policy, land planning and zoning are policy areas where regional governments usually possess a high degree of autonomy. Canada and Australia have both developed systems to permit the harmonization of national and regional goals. In Australia, the National Water Resources Act was developed on the basis of an agreed "framework of objectives", an approach also being applied to the development of a national forest strategy. An example is to be found in Canada in the form of a federal/provincial task force developing recommendations regarding the management and use of soil and water resources for agriculture.

In the United States each state is able to provide input to the development of national agricultural and environmental policies. One method of doing this is through the environmental assessment process; another is via interagency agreements between state and national agencies. Personnel exchanges, interagency committees as well as shared data bases and information exchange are common.

Belgium provides an example of a country where the pendulum of authority and initiative has swung to its three regions, which are said to be predominantly responsible for environmental problems. In most OECD countries, the administration of national parks and nature reserves is assured by the central government and often represents an area of responsibility transferred recently from agricultural to environmental ministries.

Not surprisingly, at the regional government level, the same kinds of inter-agency and departmental co-ordination problems are encountered as at the national level, and Australia cites the case of three states which have put into place their own environmental co-ordination committees. In Spain, the autonomous region of Galicia has taken a similar approach.

Another kind of structure used to integrate national and regional activities are inter-regional coordinating bodies, most often associated with the management of catchment areas and river basins. In general they represent a pragmatic effort to cut across existing, less flexible institutional lines to deal with a specific set of problems or circumstances. These play large roles in Australia, (Murray-Darling Basin Council), France (les Agences de bassin) and the United States (Chesapeake Bay project). In the United Kingdom, water authorities have considerable responsibility for water supply, pollution control, drainage and flood control. Canada and the United States have a 77 year-old International Joint Commission which deals with transboundary waters and shared resources. The Commission's approval is required for any use or alteration of transboundary waters such as the Great Lakes that affects level or natural flow.

In Canada, the Prairie Farm Rehabilitation Administration has been active for over 50 years in addressing economic development and soil management of differing soils across three prairie provinces and several major drainage basins.

Integration and co-ordination with local authorities

1.   In all OECD Member countries, local authorities must also be taken into account. It is broadly true that environmental responsibilities at the municipal level (wastewater treatment, solid waste management) tend to be devolved to these authorities, who then must be included in developing policies where wastewater, sewage sludge, or solid waste disposal become factors affecting agricultural production. Hierarchical and co-ordination problems can arise where local government, as is the case in most federal states, derive their authority from provincial and state governments as opposed to the central authority. Broad variations exist as to the degree of autonomy such local administrations exercise.

Finland provides a unique example of environmental responsibility devolved to the municipal level with the creation in 1986 of environmental committees in all municipalities with more than 3000 inhabitants responsible for protection of water and air and waste management. These committees prepare environmental protection implementation plans subsequently approved at the provincial level, and forming part of the regional and national environmental protection programmes. They provide a unique example of "bottom-up" environmental planning. The local environment committees are paralleled by municipal agricultural committees responsible for agricultural management at the local level. In the United States regional water pollution planning was undertaken in the 1970s covering point and non-point sources. Specific municipal pollution control plants must comply with these plans to qualify for federal funding.

At the opposite end of the spectrum of complex and highly developed agriculture/environment policy organs are countries in unique albeit quite different situations.

New Zealand, by virtue of her sparse population and extensive agricultural practices, has suffered little of the environmental impacts of intensive agriculture experienced in the Northern Hemisphere. Nevertheless, there have been environmental problems which require extensive co-ordination at a local level. The principal mechanisms for this are catchment authorities which develop catchment plans, administer land-use regulations and implement catchment control schemes. Each catchment authority usually contains elected representatives from local government authorities and representatives from the appropriate resource departments.

The relative absence of integrated or co-ordinated agricultural/environmental policy in Spain is attributable to entirely different circumstances. Much of Portugal's and Spain's administrative superstructure is undergoing significant change, in part due to their recent entrance to the European Community. At the national level, environmental responsiblity is spread across several departments, making co-ordination difficult. Moreover, Spain's quasi-federal system of government has resulted in most environmental responsibilities being devolved to the autonomous regions, where environmental responsibilities are often exercised by agricultural agencies. In one autonomous region, Galicia, an interdepartmental environmental commission has been created.

Greece too, by virtue of her own particular geographic and economic context, states that it has not experienced the same kind of environmental impacts common among her European neighbours, but suffers on the other hand from extensive soil erosion. A complete rationalisation of existing legislation and services has been recently carried out, and comprehensive environmental machinery has been put in place, including the Ministry of Environment, Physical Planning and Public Works responsible for overall policy, but with implementation responsibility distributed among functional departments.

Approaches to implementation: incentives versus penalties

In examining various national approaches to the implementation, monitoring and enforcement of environmental policies, one finds a continuum of "carrot and stick" approaches ranging from the purely legislative and regulatory instruments, through incentives (and disincentives) to a persuasive, voluntary approach to the resolution of environmental problems associated with agriculture. The last places considerable emphasis on public information, education, and extension, and Belgium provides the example of campaigns to involve the press in passing the environmental message.

The Scandinavian countries, The Netherlands and Turkey provide examples of a largely regulatory approach to agricultural and environmental problems. The Turkish report refers to sanctions for water pollution contained within the penal code. As a result of an increased emphasis on environmental protection, there have been discussions in Canada on including a category of "crimes against the environment" within the jurisdiction of the criminal code which is the responsibility of the federal government. The United Kingdom, on the other hand, has perhaps the most clearly enunciated policy of emphasizing voluntary approaches, with incentives for good environmental practices related to a framework of statuatory controls where these are most appropriate. The United States emphasizes a voluntary approach especially regarding the control of erosion and non-point source pollution, reinforced by disincentives against the maintenance of certain practices considered to be environmentally harmful. Failure to adopt recommended "best management practices" may disqualify those farmers participating in support programmes from receiving benefits.

Many other specific examples of both regulatory and voluntary approaches can be found in section 3 below.

Administrative organisation in some countries reflects a clear division between policy formulation and implementation. Finland, for example, possesses both Agricultural and Environment Ministries, responsible for policy within their respective areas, but with implementation, monitoring and pollution prevention delegated to National Boards, such as the National Board of Agriculture and the National Board of Waters and the Environment. In the United Kingdom, environment departments hold policy responsibility for the "physical, economic and social" environment and implementation is, for the most part, delegated to agencies such as local authorities, county and district councils, water authorities, the Nature Conservancy Council and the Countryside Commission.

Similarly, the National Agency of Environmental Protection, an agency under the Danish Ministry of the Environment, hears appeals, prepares legislation, carries out studies and evaluations, and advises regional and local authorities. The establishment of laws, however, is in the hands of the Danish Parliament. But they normally only pass framework legislation to be completed in more detail by the Minister for the Environment.

Policy making and legislative frameworks

Many OECD governments have opted for hybrid policy frameworks, typically called "strategies" setting out general policy orientation within which more specific policies, legislation and regulations can be developed, be it at the national, regional, or even local levels. Australia has adopted a National Conservation Strategy as well as National Tree and Soil Conservation programmes in collaboration with state authorities. Canada has adopted a federal policy on land use, together with Federal Land Management Principles, which are mirrored by similar policies in several provinces. Provincial agricultural land reserves or restrictions in the Canadian provinces of British Columbia, Newfoundland and Quebec, however, have an even more profound impact on regional land use.

Taking such strategies one step further, several governments including the United States, Australia and Canada have enacted umbrella-type laws, broadly falling within the category of national environmental acts. Their common feature is the requirement for environmental impact statements for proposed new activities that could have significant environmental effects. Such laws often result in the adoption at regional level of similar or complementary laws.

Spain, on the other hand, availed itself of its recent political evolution to write environmental objectives into its new constitution of 1978. Austria also amended its constitution in 1984 to oblige the federal government, all provincial governments and all local authorities, to prevent pollution.

Together the country information papers describe a wide variety of policy making and legislative frameworks. The impression they create is that, taken all together, the legislative arsenal is probably as complete as it needs to be. Emphasis now needs to be placed on implementation on the part of the various actors involved, and above all on optimum integration of agriculture and environmental considerations as new and existing policies, programmes and legislation are developed.

The role of quasi-autonomous and non-governmental bodies

The United States and United Kingdom provide examples of an implementation role for quasi-autonomous and non-governmental bodies, such as the Countryside Commission in Britain, and non-profit foundations in the United States like the Nature Conservancy. In these and other countries such as The Netherlands, private non-profit organisations, which may receive public support, are acquiring land for conservation purposes, occasionally leasing it back to farmers under certain conditions. Similarly, in Canada, the government has recently provided a grant to assist with the formation of Soil Conservation Council of Canada. This independent non-government Council, supported by agricultural and conservation individuals, organisations and government agencies, has been established with the sole purpose of working to protect Canadian soils and related water resources from loss and degradation.

In Denmark, the Danish Society for Nature Preservation has the statutory right to "raise questions" of nature conservation. In several other countries organisations and individuals can take legal action on behalf of a particular class of people to ensure that conservation issues are given full consideration during public decision making.

International policy co-ordination and integration

It is beyond the scope of this report to examine international arrangements for the co-ordination and integration of agricultural and environmental policies. Information papers from members of the European Community, however, repeatedly emphasize the role of Community policy in these areas, in particular the Common Agricultural Policy (CAP).

Community policies which bear directly on national policy approaches reflect: a) the growing importance of environmental policy at the Community level, as evidenced by the European Community Third and Fourth Action Programmes on the Environment. The Third Action programme aimed at striking a better balance between environmental policies and those of other sectors, and by the adoption of such instruments as the recent directive regarding environmental impact assessments; and b) specific new measures towards integration in the area of agriculture and the environment. The latter is illustrated by the recent agricultural structural policy adopted in March 1985, which for the first time, specifically allows measures for the protection of the environment.

The other trend would seem to be a growing tendency to acknowledge that the "hot potato" of environmental policy, historically left to individual Member states, is an area where national governments are prepared to concede a greater role to the Community. European Community initiatives in this area appear in fact to be filling certain gaps in some Member countries where environmental policy is not yet highly developed. There are currently two major sources of these initiatives. First, the Single European Act will add a section concerning the Environment to the Treaty of Rome. This will be based on the three principles of preventive action, rectifying damage at source, and "the Polluter-Pays Principle". Second, under the 4th Action Programme on the Environment (1987-1992), the Community aims to move from reacting to existing problems to a general preventive approach. The European Year of the

Environment, which started on 21 March 1987, is designed to awareness of the need for high environmental standards as a pre-requisite for a shift to preventive action.

In the Italian case, recent national measures in terms of institutions and legislation appear to have arisen at least in part from the necessity to put national policy in line with that of the Community. Policy planning and administrative organisation in Spain is being restructured in large measure to conform with Community requirements. Greece, however, has drawn attention to the considerable social costs that could follow from an abrupt, unilateral application of Community policies to its still relatively extensive agricultural sector.

Another area where the Community plays a preponderant role concerns policies directed at the reduction of agricultural surpluses. While not intended as environmental measures per se, attempts to cut back agricultural production can, in some circumstances, have the side benefit of generally reducing agricultural pressure on the environment.

A number of recent developments in the form of committee enquiries, a report to the European Parliament, and the publication of the Commission's "Green Paper" on perspectives for the Common Agricultural Policy have referred to the need to better reconcile environmental and agricultural objectives. Subsequent debates have focused on the difficulties associated with implementing policies to achieve this recognised need for integration. The Council of Europe also plays a significant role in setting environmental standards which affect european agriculture. Such recognition between environmental and other objectives are required by the Single European Act.

Another multilateral organisation having a significant impact on both agricultural and environmental policies in OECD countries is the United Nations Food and Agricultural Organisation (FAO). This agency sets international standards concerning the contamination of foodstuffs through the Codex Alimentarius and takes a leading role in pesticide control.

3.   POLICY APPROACHES TO INTEGRATION

The institutional and administrative structures described earlier are used in virtually all OECD countries to facilitate the development and implementation of policies to influence the allocation, use and appreciation of agricultural and environmental resources. Although there is consistency between the agricultural and environmental objectives of most OECD countries, specific policy objectives and the manner in which they are implemented vary considerably.

The policy instruments used to achieve agricultural and environmental objectives are usually directed towards agricultural inputs, practices, products, the capital structure of agriculture and, more recently, the enhancement of the positive contributions which agriculture can make to the environment. An integrated approach to the successful use of these instruments requires an understanding of their interaction and of the likely impact of each instrument on all policy objectives. As indicated in Section 2.0, such an approach requires extensive consultation amongst all the people involved in and likely to be affected by a policy. This section describes the instruments and mechanisms used to implement integrated policies.

Policies directed at agricultural inputs

Generally in OECD countries individual producers have freedom of choice in the selection and use of inputs, subject to restrictions only where necessary to meet product, producer and environmental safety standards. In most OECD countries the principal agricultural inputs having significant effects on the environment are agricultural pesticides and fertilizers, although hormones and specialised additives to animal feedstuffs are also important inputs for efficient production. The important positive contribution which agricultural labour can make to the maintenance of the environment is discussed in section 3.5 below.

Research, extension and advisory programmes

Farming is often conducted in a risky and uncertain environment leading to a tendency by farmers to use excessive amounts of fertilizers and pesticides in an attempt to ensure maximum yields, while minimising the risk of crop disease and pest outbreaks. Often the excessive or inappropriate application of agrochemicals results from inadequate economic appraisal and inadequate knowledge on the part of farmers of how responsive crops are to different rates of fertilizer application; how much residual phosphate and nitrate is in the soil; what the probability of pest outbreaks, etc is; and when or how much pesticide is needed to prevent such outbreaks.

In all OECD countries research, extension and technical advisory programmes are used to advise farmers about the effects of agricultural practices on the environment and ways to increase profits. Relevant research is often the first step to the more effective provision of extension and advisory services. Emphasis is placed in countries like Belgium, United States, France and Canada on ensuring that research results are quickly disseminated to farmers and are translated into practical demonstration plots at a local level. Private firms often adopt similar strategies to encourage farmers to buy minimum tillage equipment, new manure injection equipment, new agricultural chemicals, etc.

Part of the money raised from levies on inputs in countries like Austria and Sweden are used to fund research designed to increase productivity and also to develop means to assist farmers to improve the environment. In Australia levies on the sale of agricultural products are used to fund research.

As pointed out by Greece and Sweden extension activities can be very effective in disseminating the idea of the rational use of inputs to farmers. In such situations they have the dual effect of increasing net farm income and enhancing environmental quality. In Turkey, for example, the major means employed to discourage excessive fertilizer use is to provide farmers with information on the negative effects of fertilizers on the environment and the impact of this on farm income.

In Sweden, as elsewhere, many farmers have tended to over-apply fertilizers, in part due to a lack of information and in part by their failure to appreciate the difference between yield and profit. Advisory services are now being used to inform farmers of the value of farm manure as a source of nitrates and phosphates and to use commercially-produced fertilizers more efficiently. The costs and problems of applying farm manure, however, should not be underestimated.

One consequence of the over application of fertilizers common to many European countries is nitrate and phosphate run off into surface waters and leaching into ground waters. This has contributed to eutrophication problems and an increase in nitrate levels in some drinking waters above the World Health Organisation standard of 45mg/litre and the European Community standard of 50mg of nitrate per litre. The Swedish approach to this problem has been to target extension activities to the key problem recharge areas. An inventory on the application of manure and on mineral fertilizer use on each farm in the recharge area has been carried out and, where the inventory and subsequent analysis revealed that fertilizer application was excessive or where deficiencies in manure handling were observed, farmers were urged to take corrective action. A similar advisory approach to the resolution of the nitrate problem is also being taken in many other countries, particularly Belgium, Denmark, Finland, France, Germany, Norway, The Netherlands and the United States.

Many countries are also coupling intensive advisory efforts with strong public information programmes aimed at making farmers more aware of the off-site impact of their activities. Such programmes are often directed towards school children and the general media. In Austria, for example, rural secondary school children are now being offered a new subject called "area-planning and environmental protection". Experience with these programmes suggests that farmers are willing to take a voluntary approach to the better management of agricultural inputs, particularly when correction of the problem is also likely to increase net farm income.

Belgium and the United Kingdom emphasize the effectiveness of a voluntary approach to reduce the negative impact and enhance the positive impact of agriculture on the environment. Importantly, such voluntary programmes also involve farmers in the process of designing and developing extension and advisory programmes and are frequently dependent on integrated research programmes for their information.

A notable recent development, reported in many country information papers, has been the adaptation of extension and advisory programmes to encourage farmers to use integrated pest management procedures.

Quantity restrictions

In places where nitrate pollution of ground water is a particularly serious problem several countries are planning to restrict mineral fertilizer use. At Laholm in Sweden, for example, applications around the re-charge area for a municipal well were restricted to 100 kg nitrate per hectare per year in 1975. In less than 2 years nitrate concentrations in this well fell from 80 mg to 40 mg of nitrate per litre. In subsequent years they continued to fall and since 1979 they have remained below 30 mg of nitrate per litre.

In other countries such as Austria and Germany, the spreading of fertilizers in some water catchment areas is similarly restricted or prohibited. Recently, the German Parliament amended its federal Water Management Act to provide for the designation by the laender of water protection areas so that the use of mineral fertilizers and other related practices can be either restricted or prohibited. Under these amendments Laender may compensate farmers who suffer economic hardship as a result of these restrictions. In at least one Land a levy on downstream water users is being discussed to finance such compensation. A similar approach is also being pilot tested in the Loire Valley in France.

In other countries, such as the United Kingdom and Denmark, some farmers are advocating the introduction of fertilizer quotas as a means of reducing agricultural surpluses and the pollution of ground and surface waters.

Input subsidies, charges and levies

In Finland, Sweden and Austria price regulation charges, input taxes and input levies have recently been introduced for the multiple purposes of a) financing the cost of exporting grain; b) reducing the use of pesticides and fertilizers; and c) providing funds for research and extension. In Sweden a 20 per cent price regulation charge and a 5 per cent input tax are levied on the value of the nitrogen and phosphorous in all fertilizers and the value of each unit of active ingredient in all pesticides. The charge on fertilizers was first introduced in the 1970s and then gradually increased until 1984. The charge and tax on pesticides was introduced in 1986 following extensive debate on the adverse effects of pesticides on the environment.

In Austria, a market regulation was introduced in 1986 to finance soil protection measures; to promote the development of crops other than wheat; to reduce the use of fertilizers; and to finance the sale of surplus grain through a tax or levy on active ingredients. In Austria, these taxes are known as a "Fertilizer Tax" or "Soil Protection Tax". A similar situation exists in Finland where producers are required to contribute, through a fertilizer tax, to the cost of exporting agricultural produce. All the above countries have noticed or are expecting a substantial decline in pesticide and fertilizer use as a result of the introduction of these charges.

Similar changes have also been noticed in New Zealand where, until recently, the use of farm fertilizers was subsidised. The New Zealand government has now decided that all sectors of its economy should be exposed to market forces and, within the agricultural sector, it has withdrawn all fertilizer subsidies, product price supports, all land development encouragement loans and all tax concessions. As a result New Zealand predicts that its agriculture will become a) more extensive and b) more diversified. Phosphate fertilizer use is expected to decline by 45 per cent in the year beginning June 1986 as a result of the elimination of fertilizer subsidies and price support.

A characteristic of each of these taxes, charges and approaches in Austria, Finland, New Zealand and Sweden is that, although not primarily introduced to overcome environmental problems, each has produced substantial environmental benefits by decreasing fertilizer and pesticide use. This decrease has not only resulted from the higher cost of these inputs, which in turn encourages farmers to avoid over-application but also because it has made them more aware of the potential for extra profits by managing them with greater care. The programmes have changed farmers perceptions of fertilizers, and led to a recognition that profit maximisation and production maximisation may require different quantities of inputs.

A different approach to those described above can be found in almost all German laender where, to protect native flora and fauna, farmers are paid to leave pre-determined crop edges unsprayed when they apply pesticides. Eight out of eleven laender also offer "grassland extensification schemes" which offer payments to farmers who refrain from spraying pesticides, reduce levels of fertilizer or leave meadows unused during main insect hatching periods.

In The Netherlands, to reduce nitrate and phosphate pollution and, also, the acidification problems associated with intensive animal husbandry in 1988 a general levy and a surplus levy will be introduced. The general levy applies to all livestock feed manufacturers and will be used to meet research costs associated with manure surpluses and advisory services. The surplus levy is in proportion to amount of surplus animal manure expressed in terms of the kilograms of phosphate which can be expected to be produced each year. Farmers will have to keep manure accounts and those who own sufficient land to spread manure without exceeding predetermined maximums are exempt from paying the surplus levy. In line with the "Polluter-Pays Principle" the surplus levy is designed to cover the costs of maintaining manure banks; and meeting the infrastructure costs of removing, treating and disposing of surplus animal manure.

In Australia, the United States and New Zealand water for irrigation is generally either subsidised or free. Generally, in Australia, farmers have not been required to pay the full capital costs of water supply. Low water charges have caused land to be used more intensively and more water to be used per hectare than would be the case if it were priced at its true economic cost. Cheaply priced water means that there has been no incentive to adopt technologies such as land forming which, through the more efficient use of water, would reduce salinity. Having recognised this, Australia is now reviewing its water pricing policies. One state, New South Wales, has substantially increased its water charges and other states are expected to follow. The result was the almost immediate widespread adoption of more efficient farming practices and a decline in salinity levels.

In several countries there is now an interesting debate concerning the merits of a combination of a fertilizer tax and a fertilizer quota. Quotas are often administratively more expensive to implement than price based measures as the system requires records to be kept for many farmers rather than a few dealers. Nevertheless they tend to have less impact on farm income than measures affecting input prices. From the environmental vantage point, quotas are not sensitive to price fluctuations and hence can be relied upon to reduce fertilizer use. The use of taxes to limit fertiliser use can also be readily overwhelmed by commercial price changes. For example, the price of West European ammonium sulphate fell by 43 per cent between September 1985 and September 1986. On the other hand, quotas tend to create locational inefficiencies unless they are transferable. At the time of writing, no national quota schemes for agricultural fertilizers or pesticides exist, although in effect input quotas for water have existed in many countries such as the western parts of the United States, Australia, Spain and Portugal. These "quotas" are usually known as water rights and in recent years several countries such as Australia and the United States have begun to make these rights transferable.

Another problem associated with the under-pricing of inputs has also been recently recognised in Australia. This is the payment of fodder subsidies during periods of drought, which encourages the retention of livestock on farms by lowering the cost of feed which makes it less attractive to destock and, in turn, increases the probability of both pasture and land degradation. It also discourages farmers from acting to avoid the negative environmental consequences of drought, like soil erosion, which tend in the long run to decrease potential productivity. Recent official recognition of the adverse consequences of input subsidies of this kind has led to a policy review recommending that drought policy should encourage the more efficient use of resources. Emphasis is now to be given to making resource maintenance a high priority for farmers. The continuation of drought fodder subsidies was not recommended by this Committee and will not be continued. These recommendations have largely been endorsed by the Australian Commonwealth and state governments.

Tariff policies, preferential trading agreements and product price support policies may also affect the use of inputs and intensity of agricultural production. For example, in The Netherlands, part of the increase in intensive animal husbandry has been stimulated by price support for animal products within the European Community and the duty free or low tariff rate applied to many stock-feed ingredients from developing countries. Conversely, Switzerland places a tariff on feed inputs to encourage Swiss farmers to grow their own forage, thereby encouraging on one hand partial self-sufficiency and, on the other hand, an acceptable balance between national forage and animal production.

Generally the experience with varying the price of agricultural inputs so that they take account of the environmental costs of agriculture has been favourable. Sweden, Finland, Austria, New Zealand, Portugal and Australia have all found that increasing the cost of agricultural inputs makes farmers more aware of the additional profits which can be made by the more efficient use of inputs and also, reduces both on and off farm pollution by internalising some of the off-site costs of pollution from agriculture. Where input prices do not reflect their true on and off site costs, the experience described above suggests that appropriate regulations and standards may be needed to prevent environmental problems.

Input standards and regulations

Many agricultural inputs contain chemicals which can have unintended side effects on human health and flora and fauna as well as the long-term productivity of agricultural land. Consequently it is standard practice in OECD countries to prohibit the manufacture, sale and use of agricultural inputs which do not meet specified standards or have not undergone appropriate product tests. In recent years, following the development of analytical methods to detect trace elements, people have begun to express concern over heavy metal accumulation in soils and pesticide residues in water. For example, it has been found in Sweden that the present rate of increase of

101

cadmium in soil is 0.5 per cent per annum. In Germany, the Council of Environmental Advisors to the Government has observed that the present cadmium burden from mineral fertilizers is around 3-5 g/ha. These experts also observed that the cadmium burden originating in animal manure and sewage sludge is generally higher, albeit in most cases less widespread, than that associated with mineral fertilizers.

Cadmium and other heavy metals are common in sewage sludge, some mineral fertilizers and also some feed additives. Portugal, for example, reports that it is experiencing environmental problems as a result of the existence of copper and zinc in the feed supplements used for livestock breeding. Some, but not all OECD countries, have set limits on the maximum amount of cadmium which may remain in mineral fertilizer sold to farmers. The Netherlands considers that an international approach is needed to reduce cadmium pollution and has proposed to the European Community that a community wide standard should be introduced for feedstuffs and compound feeds. In the interim, the government has entered into a two-year agreement with the Commodity Board for Industry to keep cadmium levels below specified levels. A monitoring programme has confirmed that the industry is meeting the standards set for compound feeds, but that the standard for normal feedstuffs is often exceeded.

Denmark is preparing new limits for cadmium in phosphorous fertilizers. A transition arrangement is foreseen in which the maximum permissible amount of cadmium will be gradually reduced over a number of years to enable the industry to adjust. In Sweden the cadmium standards for mineral fertilizers are being strengthened and within the European Community, research into ways of further reducing the cadmium content of mineral fertilizers is being conducted in co-operation with industry. Cadmium standards for sewage sludge also exist at the level of the European Community and also in countries like Austria, Japan, Sweden and the United States.

In parallel with these developments, countries are also strengthening controls to reduce the impact on agriculture of pollution from other sources. The German Council of Environmental Advisors, for example, make the point that pollution from industrial air pollution is increasing the cadmium content of German soils at a similar order of magnitude as mineral fertilizers. Similar experiences have been reported in Japan for cadmium, copper and arsenic. Water pollution from industrial sources and the spreading of sewage sludge is also having a substantial impact on agriculture in some areas.

In the last three decades there has been a dramatic increase in the use of fertilizers and pesticides. Most of the increase in fertilizer use occured during the 1950s and 1960s. For example, between 1951 and 1970 the consumption of nitrogenous fertilizers more than doubled in Belgium and The Netherlands; tripled in Germany, Luxembourg and Italy; quadrupled in the United Kingdom and underwent a five-fold increase in France. From 1970 to 1984 the average increase in the use of nitrogenous fertilizers was 56 per cent, although the increase in the use of phosphate fertilizers was only 1 per cent. The country information papers indicate that:

-- in the United States since 1964 pesticide use has doubled from 540 million tons of active ingredients to 1.1 billion tons, with the most significant growth being in herbicides;

-- in Denmark from 1950 to 1984 the use of pesticides increased by a factor of 5; and

-- in Norway, Sweden and the United States pesticide residues have been found in samples of both ground and surface water in sufficient quantities to justify nationwide surveys of the extent of this pollution.

In recent years there has also been a shift towards the use of low dosage and less persistent pesticides requiring more frequent application. In Denmark between 1981 and 1984, for example, while the quantity of pesticides used in agriculture increased by 25 per cent the average number of applications of pesticide per year increased 115 per cent. Though intended to minimise the impact on the environment, the implications of these new developments for the survival and diversity of flora and fauna have yet to be fully assessed. Research in several countries, however, is examining this issue.

In nearly all OECD countries the sale of agricultural pesticides and herbicides is now subject to legislation which requires extensive testing and evaluation of any new products before a government will license their sale for a specified use. In the United States these requirements, enforced by the Environment Protection Agency, generally add about 1.8 per cent to the cost of pesticides. Several countries, including Sweden, have begun to re-evaluate pesticides which were registered during a period when different registration criteria applied. Sweden plans to complete this re-evaluation by 1990.

Harmonization of standards for pesticides in OECD countries is as of yet incomplete. For example, Belgium generally licenses pesticides for 10 years, while Norway only licenses them for 5 years and Japan for 3 years. Moreover, chemicals banned in some countries are still used in others. For example, the use of DDT for forestry purposes is banned in most countries, but still permitted in Norway and Turkey. Extensive but differing labelling standards also exist in most countries. Table 3.1 identifies the normal period for registration in OECD countries and whether or not countries have adopted legislation which restricts the use of pesticides. All countries which issue licences for an indefinite period issue them on the understanding that they are subject to immediate withdrawal should information which questions their safety becomes available.

Most countries reserve the right to withdraw licences to sell pesticides if new data indicates that a product may no longer meet safety standards. As France points out, the suppression of such products usually stimulates research to find substitutes and, in the long run, the impact on farm income is usually minimal. In the United States, a chemical suspected of having adverse side effects on human health or the environment becomes the object of a Special Review including an extensive risk/benefit analysis. This was the case, for example, for the insecticide Chlorobenzilate. Following a Special Review, its registration was withdrawn for most uses except on citrus. The review imposed a requirement for a label statement, restricting it for use in conjunction with an integrated pest management programme.

Table 3.1

NORMAL PERIOD OF PESTICIDE REGISTRATION BY OECD COUNTRY (1)

Country	Normal period for pesticide licence (2)
Australia	1-3 years (3)
Austria	Indefinite (4)
Belgium	10 years
Canada	5 years
Denmark	Indefinite (5)
Finland	5 years
France	10 years
Germany	10 years
Greece	5 years
Italy	Indefinite
Japan	3 years
Netherlands	5 years
New Zealand	Indefinite
Norway	5 years
Portugal	10 years
Spain	5 years
Sweden	5 years
Switzerland	Indefinite
Turkey	About 3 years
United Kingdom	10 years
United States	5 years

1. The use and registration of pesticides are covered by Act of Parliament in all countries.

2. Some countries register rather than license pesticides for use.

3. One year in two states and 3 years in all others.

4. New legislation being prepared with a 10 year period.

5. New legislation is being considered by Parliament with a proposed 8 year period.

There is also a need to consider the perverse effects of pesticide regulations. For example, until recently, the Environmental Protection Agency in the United States prohibited the use of pesticides in crop rotations unless data were provided to demonstrate that residues in rotational crops would not occur. The unintended outcome was the suppression of otherwise favourable crop rotation systems because the chemical industry chose not to generate the data. Recent policy changes permit finite maximum residue levels to be legally established for follow-on crops without such data.

In order to further reduce the ecological risks arising from the use of pesticides a new German Plant Protection Act was introduced in 1987. It includes, amongst other things, provisions to protect groundwater, a legal obligation for users to acquire the necessary expert application skills and a prohibition of the use of pesticides in non-agricultural areas. Similarly, to protect people from drinking water polluted from agricultural and other sources the European Community has established a maximum permitted concentration in drinking water of 0.1ug/1 for individual pesticides, organochlorines, fungicides and herbicides. There is also an overall maximum permitted concentration for all pesticides of 0.5ug/1 in drinking water. Similarly, since 1980 the European Community has laid down standards on the quality of water intended for human consumption. The maximum concentration of nitrates allowed in drinking water is 50 mg/litre.

Generally, the information provided by Member countries confirms that sophisticated review procedures are now in place in most countries for all new pesticides. There are, however, a significant number of existing pesticides which have been in use for many years and which have not been subject to the same review procedures. Some of this latter group of pesticides are now being subject to data call in programmes. In the United States, for example, pesticide law permits the Environmental Protection Agency to suspend products unless the registered manufacturer agrees to supply data necessary to meet current requirements. A data call-in programme fully identifies the required information. A registrant has 90 days to commit himself to the production of the required information on a specified schedule. If he does not commit or does not meet the agreed schedule, the product may be suspended from sale.

Integrated input programmes

New integrated programmes have recently been developed in Sweden and Denmark to address all of the above issues simultaneously. Initially, both governments set a firm policy target to reduce national pesticide use to a specified level and subsequently developed programmes to achieve these targets.

In 1985 the Swedish Government announced that it would introduce measures to reduce the use of pesticides in agriculture by 50 per cent over 5 years. The Danish Government announced in 1986 that they would initially reduce the consumption of pesticides by 25 per cent within 3 years and set a goal to reduce current pesticide use by a total of 50 per cent within 10 years. The Swedish proposals which are still under discussion include:

-- stricter review procedures, including demonstration of a need for the product;

-- a requirement that products be tested at three different doses, two of which must be lower than that necessary for total eradication; and

-- preventing the use of chemicals when other equally economical methods of control are available.

In the context of the ongoing discussions described above the Swedish government has already decided that:

-- all farmers will be required to attend an approved 3-day training course before being allowed to purchase agricultural pesticides; and

-- it will introduce a regular and compulsory equipment testing for all farm machinery.

Importantly, these proposals are supported by the <u>Swedish</u> agricultural organisations which helped develop them. The integrated input programme described above is to be mainly financed by the 5 per cent input tax on fertilizers and pesticides.

The <u>Danish</u> Government is still developing the measures that it will adopt to meet its 30 per cent pesticide reduction target over 3 years. Measures being considered include:

-- improved early warning systems coupled with a more effective system of assessing economic risks;

-- the prohibition of the use of pesticides in areas of special environmental interest;

-- the introduction of pesticide exclusion strips;

-- the introduction of high pesticide charges;

-- subsidising cultivation methods based on the reduced use of pesticides; and

-- more research on alternative production practices.

The <u>Danish</u> Government has also recently released an action plan to help control the pollution of water by reducing nitrogen discharges by 50 per cent and phosphorous discharges by 80 per cent within 3 years. The plan proposes actions to reduce discharge from municipal water treatment plants, industry and agriculture. In the agricultural sector the Danish Government has now decided to:

-- require farmers with more than 30 animal units to have at least 9 months manure storage capacity before 1993;

-- increase subsidies for the expansion of manure storage facilities;

-- improve advisory and extension services on application rates and techniques;

-- require all animal manure to be ploughed in within 12 hours of spreading;

-- require all farmers from 1988 to prepare fertilizer and animal manure application plans which are available to the authorities;

-- require each farmer to establish 45 per cent of their farm under green fields which contain either a catch crop, ploughed in stubble or grassland in 1988, and then increase this proportion to 55 per cent in 1989 and 65 per cent in 1990; and

-- strengthen guidelines for freshwater fish farms.

In The Netherlands a memorandum on pesticides and crop protection techniques has been designed to reduce pesticide use by:

-- requiring advisory officers to give extra attention to the promotion alternative biological methods of pest control and, also, the improvement of farm hygiene and cultivation methods so that less pesticides are needed;

-- establishing thresholds below which diseases and pests can be tolerated without economic damage;

-- establishing early warning systems so that farmers need only apply pesticides when they are needed; and

-- developing more selective chemicals and more confined application methods.

In OECD countries, farmers generally perceive it as being important and in their interest to safeguard the long-term productivity and potential of the land they use. Consequently, the need for restrictions on agricultural practices is perceived by some countries as being necessary only when the off-farm costs of these practices are significant. Economic pressures, international market competition and other factors can over-ride this general philosophy of sustainable use and development with the consequence that regulations are also being introduced to maintain and protect long-term productivity. This is particularly common in the extensive rangeland areas found in Australia, the United States and Canada.

In the case of non-point sources of pollution, however it is difficult and at times impossible to identify the farmers who are responsible for causing it. Research in Sweden, Denmark and the United States has indicated that because the rate of water run-off and soil leaching also determines the quantity of pollutants which leave a farm, it is not necessarily those farmers who use the most inputs who impose the greatest costs on others. All OECD countries have developed extensive extension and advisory services to encourage farmers to improve their management practices and, as far a possible, conduct them in a way which enhances the quality of the environment.

Policies directed at agricultural practices

Research, extension and advisory programmes

In recent years many governments have re-oriented their research, extension and technical advisory programmes to give greater emphasis to the environmental consequences and risks associated with different agricultural practices. Examples of this evolution include:

-- the identification and promotion of best management practices throughout the United States;

-- the development of codes of practice in Australia;

-- the modification of the syllabuses taught at university and other training institutions to ensure that graduates are prepared and able to explain the environmental consequences of agricultural practices, in countries such as Switzerland, France, Germany, and Sweden;

-- the development of advisory services for agriculture which reflect environmental concerns in the United Kingdom;

-- the provision of specialised advisory services for soil conservation in Canada; and

-- the establishment of an inter-disciplinary environmental group within the Danish consultancy service system.

In Australia, Canada and Switzerland, governments also have introduced special funding arrangements to promote the development at state, provincial and canton level of integrated research, extension and technical advisory programmes in these areas. For example, within the Canadian Government's Economic and Regional Development Agreement Framework, natural resource initiatives under agricultural subsidy agreements may have funds specifically allocated to demonstration projects, conservation research and the provision of technical advice relating to soil and water management. Similarly, in Australia, under the National Soil Conservation Programme substantial grants are made available to States for specific soil conservation research programmes.

Usually, extension and technical advisory services are provided to farmers either free of charge or at rates substantially less than their true cost. Some countries are beginning to charge for advisory services. For example, in Finland, extension and advisory services are provided through farmer associations with only 50 per cent funding from government. Similar arrangements exist in Sweden. After 1987 charges will also be made for certain extension services in England and Wales but advice on environmental matters will generally remain free. New Zealand and at least one Australian State are also considering the desirability of charging for advisory services. The practice of charging for analytical services such as soil analysis is widespread. To increase the use of such services in the French departments of Nord and Pas-de-Calais the innovative approach of providing free nitrate analyses at the end of winter for a restricted period is also being pilot tested. It is hoped that via this method farmers will recognise the financial and environmental benefits of these services and, once charges are reintroduced, continue to use them.

As illustrated by the case of Belgium, many countries find it difficult to effectively evaluate extension and advisory services. There can also be problems in ensuring that different advisory activities are integrated so that farmers are not receiving contradictory messages. This has led in recent years to a rationalisation of extension services so that most effort is being directed to priority problem areas. The recent targeting and rearrangement of soil erosion advisory activities within the Soil Conservation Service of the United States has increased the amount of effort spent on areas with severe soil erosion problems and decreased it in areas where the problem is less significant. As a result of these and other activities, soil erosion control has become more efficient in terms of effort per ton of soil saved. They have found that policies and advisory activities should be targeted on the encouragement of practices which provide maximum environmental benefit.

Another recent development has been the provision by specialised advisory services of additional information to enable farmers to adopt more efficient practices. This includes the growing use of computer technology, enabling farmers for example to:

-- adopt more effective integrated pest management strategies on the basis of accurate information on the probability of a pest outbreak and the presence of natural predators;

-- obtain more precise information on the quality of animal manure, nitrate deficiency in soils, etc; and

-- obtain more reliable market forecasts.

One example of this is the French Minitel videotex system which enables farmers to consult any data bank of their choice at minimal cost via telephone. Slightly different but reportedly equally effective schemes have also been established in Australia, Finland and in Italy with its "Agrovideotel" system. The United States Soil Conservation Service is installing a computer network in its 3,000 offices and developing software programmes to provide producers with access to information on the latest technology. It is anticipated this will assist farmers to minimise adverse impacts on the environment and to make environmental improvements where possible within their management plans. Private advisory and consultancy services, some run by chemical companies, are also beginning to provide similar services to their clients. Farm advisory and consultancy groups are also well established in many countries.

Economic Incentives

Much advisory and extension activity is devoted to promoting the adoption of new or alternative technologies. As Canada has found from experience, however, where high costs are associated with a change in an agricultural practice the probability of adoption is significantly reduced. This is frequently the case where substantial environmental gains but few economic benefits are to be derived from the change. In such situations economic incentives for the adoption of a new technology are often justified because society stands to reap significant environmental benefits from the change. As the Canadian report noted, "further progress may entail the short-term use of public assistance to help bridge the gap between conservation investments and conservation returns". Examples of such an approach include:

-- grants for the construction of soil contour banks in Australia;

-- the supply of inputs such as trees at reduced prices in Turkey;

-- sharing with farmers the costs of introducing of conservation practices to reduce erosion and flood damage and to improve wildlife habitat in Canada;

-- the provision of grants and the implementation of other cost sharing procedures which encourage farmers to improve wildlife habitat;

-- European Community grants to encourage capital investments and farming practices which protect and improve the environment;

-- provisions for the accelerated depreciation of equipment essential to a new practice and tax write-offs in the year of expenditure for soil erosion prevention and water conservation works in Australia; and

-- grants for the construction of more efficient manure storage facilities in The Netherlands, Sweden and Denmark.

Such schemes, especially when offered only for a limited period, often gain additional impetus from the private sector which is usually quick to promote the change via the sale of new machinery, new chemicals, etc. Another example is the Dutch Development and Reconstruction fund which, amongst other things, is being used to stimulate the development of techniques to reduce the production of animal manure as well as methods for processing and marketing animal manure. Grants are only given to cover the direct cost of setting up a project. Preference is given to projects which increase the quality of the manure and also those which improve its export potential.

One alternative approach to the use of economic incentives was used in Portugal between 1965 and 1974 to encourage farmers to adopt cropping systems which only use soil within capability. Long term, low interest rate loans, with up to a 20 per cent grant, were offered to farmers who agreed to comply with conservation objectives and only use their land according to its capability. Both tenants and landholders could qualify and the programme applied to almost all crops except a few like rice and grapes. In 1974 following a political change this early cross-compliance programme was abandoned.

Regulations

In cases where the potential off-site costs of inappropriate agricultural practices are significant it is usual for governments to either regulate or prohibit them. In Europe two practices receiving considerable attention are straw and stubble burning and the spreading of manure and sewage sludge. In the United Kingdom, local authorities have powers to prohibit and regulate straw burning. Straw burning will be banned after 1989 in Denmark. In the United States air pollution control laws similarly restrict the burning of agricultural residues and, in some parts of Spain, permission is required to burn stubble.

To reduce the run-off and leaching of nitrates, phosphates and heavy metals, the spreading of manure and sludge on frozen ground is prohibited throughout most of northern Europe. In Denmark liquid manure must be ploughed or harrowed into the soil within 24 hours unless it is applied to a crop or pasture. Similar provisions apply in some Austrian and German Laender. In Finland the National Board of Water and Environment has recommended that farmers should not spread manure close to a water course. The safe distance is considered to be more than 20 - 50 metres depending on circumstances.

In The Netherlands, a number of regulatory measures have been introduced to reduce nitrate, phosphate and heavy metal pollution; reduce acidifying effects of ammonia on soil; and reduce the contribution of animal manure to acid deposition. These measures complement the economic measures described earlier and include the specification of maximum dressings expressed in terms of the volume of phosphate in the manure and the type of land use. These limits on the maximum amount of animal manure which may be applied to an area of land have led to the development of a manure bank and a growing market for animal manure. The maxima are different for grass land and crop land and will be reduced by approximately 20 to 30 per cent in 1991 and a further 12 to 30 per cent in 1995. The final maxima will then be determined around 2000. As part of the enforcement programme, farmers are required to prepare and keep records of the amount of manure they produce and spread. Different standards apply for phosphate-saturated lands and nature conservation areas. Spreading is only permitted during the growing season and must be incorporated within one day of spreading. Some companies have responded to these regulations by developing and promoting the use of liquid manure injection machines.

Many other examples of restrictions on certain agricultural practices exist throughout all OECD countries, including:

-- the prevention of cultivation of highly erodible soils in some parts of Australia;

-- the prohibition of tillage within a specified distance of a water course;

-- the requirement for a licence before any form of waste from a large animal handling facility is disposed of into a water course in the United States; and

-- the prevention of sludge spreading on land with more than 200g of cadmium per hectare in Denmark.

In the area of pesticide application many regulations and standards apply. For example, in Sweden the aerial application of pesticides is prohibited and in Greece it is only permitted under a special licence for the control of the pest "Daucus Leae" with restrictions on time of use, dosage and chemical used. In Belgium, restrictions on pesticide application practices are also used to protect wildlife. For example, pesticides which are toxic to bees may not be applied during the blossom season and to protect birds and small game certain weed-killers may not be applied after 20th April.

A requirement that operators be licensed to purchase and apply certain chemicals are also becoming more common among OECD countries such as Sweden and the United Kingdom. In the United States, since 1974, the use of some toxic pesticides requires a certificate and in California permits are required to apply certain pesticides. This State has recently changed its liability laws with a view to making the users of certain listed carcogenic and reproductively toxic pesticides responsible for the consequences of using them. Few pesticides have been listed but, if widened to apply to a significant number, this approach of changing liability laws could have a substantial impact on pesticide application and other agricultural practices. Given the difficulties in policing end use in Australia, however, the control of pesticides still generally occurs at the point of manufacture and sale.

Countries such as Spain report and FAO statistics confirm that accidents still do occur in the application of chemicals. Spain, for example, reports an unusual case involving the death of 20,000 birds in the biological reservoir of Donana, attributed to the excessive use of an unidentified toxic pesticide. This disaster shocked public opinion with the consequence that more severe controls will come into effect. In Italy, physicians are required by law to report cases of pesticide poisoning.

An interesting recent development in several countries has been the introduction at a regional level of "right to farm legislation". Such legislation addresses a problem common to peri-urban areas where people seek to reside in the tranquility of a rural environment undisturbed by farm noise and odour. Complaints and calls for the prohibition of certain agricultural practices has led to the passage of legislation in Canada and the United States permitting farmers to continue with traditional practices, subject to some guidelines and restrictions.

Policies directed at agricultural products

Over the last decade, agricultural production in most OECD countries has been characterised by substantial structural change and the adoption of efficient technology linked to a greater degree of intensification which, in some cases, has led to greater environmental problems. Standards for the sale of products have also been strengthened and, in some countries, there has been a growing realisation of the need to adjust price arrangements so as to encourage the adoption of environmentally favourable production systems while meeting consumers' needs and wishes.

Product standards and marketing arrangements

Product standards and marketing arrangements to protect human health exist in all OECD countries. For example, most countries prohibit the spraying of pesticides on certain vegetable products within a specified number of days prior to sale. Similarly, maximum permissible levels of pesticide residues which may remain in feedstuffs at the time of sale are usually specified. Standards and regulations on the maximum residue of hormones and antibiotics also exist for animal products. These standards indirectly influence the degree and timing of the use of the products in agriculture and hence, the quality of the environment. They also have an impact on pesticide research as companies develop new products.

Marketing standards also affect the use of agricultural pesticides. There is, for example, a strong market demand for blemish-free fruit and, in response, countries have developed different grades for fruit and vegetable products. Such grading encourages farmers to apply pesticides and other inputs to ensure a blemishfree crop. A change in these standards could influence the quantity of pesticides used. Consequently, in the United States, a review of marketing standards and grades is required under the National Environmental Protection Act (NEPA) to consider their impact on the environment in cases where a change in grading standard could have a significant environmental effect. Integrated pest management schemes, in particular, may be feasible if the grading standard for premium grade fruit is set at one level but may prove unsatisfactory if it is set at another.

Private food processing firms may also place restrictions on the type of pesticides which may be used in products sold to them. In particular, some American canned food producers will not purchase agricultural products from farmers who use any pesticide under review. Crop insurance schemes covering pest damage can also be effective in reducing the need for chemicals.

Recently, there has been a growing demand in several countries for organically grown produce, or health food. Several countries believe the further development of this trend to be particularly attractive from an environmental viewpoint. Examples include the development of markets for free-range eggs and chickens and organically grown food in countries like France, Switzerland and New Zealand. Austria has issued a set of guidelines defining which products may be sold as being organically or biologically produced. These guidelines include restrictions on production locations relative to highways; prohibit the use of mineral fertilizers, pesticides or sewage sludge; and limit animal density to the equivalent of 2.5 cows or 15 pigs per hectare.

As a result of the development of markets for organically grown food, the need to develop guidelines on international standards for organically grown products so that no country obtains an unfair trading advantage is being discussed among several countries. Extension programmes and training courses are being adapted in several countries to encourage the production of organically grown produce. The European Community is developing regulations for the monitoring, production, and labelling of organically grown products. It is also offering financial assistance to encourage the further development of this industry.

Product price support schemes and levies

Product prices, in combination with input costs, climatic conditions and natural resources determine the location, intensity and nature of agricultural production. Two different approaches to the use of product price arrangements are presented in the country information papers. The first approach, described in the New Zealand paper, is to expose all sectors of the economy to market forces in the expectation that this will create a more efficient agricultural sector and, in turn, lead to a reduction in pressure on the environment as agriculture becomes a) more extensive; and b) more diversified. This approach also recognises the dependence of production intensity and crop yields on price levels. As pointed out by countries such as Australia, The Netherlands, New Zealand and Greece, in the long run, the intensity of production increases as the price received by farmers increases and decreases when the price decreases. The second approach described in the papers begins with the notion that product price support arrangements are necessary to ensure that farmers receive adequate incomes, that regional economies are developed and that self sufficiency in key food items is achieved. These arguments are expressed in several of the papers from members of the European Community, Switzerland, Norway and Finland.

Except insofar as they have a bearing on environmental questions, it is not the purpose of this report to address these approaches. It seems fair to state, however, that until very recently, while a wide range of political, economic and social considerations have been taken into account in formulating the above-mentioned price policies, none have been formulated in response to, or principally in response to an environmental issue.

As pointed out by Norway, past product price support arrangements have often been introduced without consideration of their likely effects on the pollution of the environment. Greece makes the point that the promotion of exports through price support arrangements has led to the intensification of agriculture and the creation of surpluses, which in turn have led to environmental deterioration. Denmark, recognising this, suggests that there is a need to consider the extent to which product price support schemes distort the structure of agriculture and create unwanted production problems. In recognition of this problem many countries, particularly those within the European Community, are now seeking ways to promote the extensification of agriculture as a means to reduce surpluses and improve the environment.

Generally programmes which reduce surplus production, if properly targeted, can have substantial benefits for the environment. Examples of such schemes include the Austrian, Finnish and Swedish approach of using input taxes and charges to finance the sale of surplus grain. Co-responsibility levies such as those which exist within the European Community and countries like Sweden and Austria can also be targeted to help ensure that self-sufficiency goals are achieved without compromising environmental objectives. Price schemes which offer a lower price for over quota production of milk such as those in the European Community and Finland can also have beneficial effects. If these quotas are not transferable, however, they can prevent structural change by preventing agricultural production from shifting to regions where it has less environmental impact. The merits of allowing such a shift need to be considered against the impact of structural change on regional economies. Consideration also needs to be given to the land use that is likely to occur as a result of the change.

Several countries, such as Sweden and France are concerned about the effects on the environment of land reverting from extensive crop production to grazing land or, alternatively, being withdrawn from production. Others are implementing a variety of schemes which, in effect, pay people to take land out of production. Typically, these schemes operate amongst economically depressed industries and in economically disadvantaged regions. Often they have both welfare or income support and environmental objectives. For example:

-- in Finland, the government contracts with farmers to leave fields fallow for one year;

-- in Germany, one laender launched a trial green fallow programme in 1986 which offers farmers a premium of between 1 000 and 1 200 DM/ha, depending on soil type, for leaving up to 20 per cent of their land fallow for one year;

-- in Sweden, a scheme will be introduced for 1987 which will pay farmers who place at least 10 per cent of their arable land in summer fallow in 1987 and then sow an autumn catch crop, between SKr 100 and SKr 2,400 per hectare, in proportion to the potential productivity of their land;

-- payments are also available in Sweden for farmers who wish to decrease cattle production;

-- in Spain, as a result of its entry into the European Community, farmers are being paid to pull up vineyards, with 50 per cent of the necessary funds coming from the government and 50 per cent from the European Community; and

-- in Portugal the re-afforestation of non-agricultural lands is subsidized by up to 70 per cent.

Italy is now evaluating the possibility of changing intensive crop production farms to extensive grazing farms through the use of payments which complement farmers' incomes. Australia has and is continuing to use similar payments to farmers to pull out unprofitable fruit trees and grape vines.

In the past, several similar schemes have been tried in the United States albeit with varying degrees of success. Building on these experiences, the American Government in the 1985 Food Security Act established legislation which:

-- invites bids from farmers to enter into a 10-year agreement which, in return for an annual rental payment, places highly erodible crop land into a conservation reserve subject to an approved conservation plan, and which may not be grazed, layed-off or cropped;

-- requires farmers, who crop highly erodible lands and wish to qualify for price support payments after 1990, to prepare and comply with an approved conservation plan; and

-- provides that any farmer who drains and crops a wetland or brings highly erodible lands into crop production in the absence of an approved conservation plan will not be eligible to participate in the majority of government assistance programmes.

Four principal concepts underline this innovative approach. First, surplus reduction is specifically "targeted" to one of America's principal environmental problems. Second, the programme has a 10-year time frame which means that there is reason to expect that there will be time for a substantial habitat change with significant benefits to flora and fauna. Third, the programme firmly establishes the principle of "cross-compliance": that is, if society is going to offer above world market prices to producers then, in return, it has a right to require them to adopt environmentally acceptable agricultural practices. Fourth, it is designed to prevent the tendency of price support programmes to expand and intensify agricultural production in marginal areas. A companion provision of the law will cause a reduction in price support over time to minimise the tendency of this programme to maintain the intensity of agricultural production across all other areas. One state, Minnesota, has introduced a complementary programme which provides for both 10 year and permanent conservation easements in return for the payment of either a lump sum or four equal consecutive payments. This State considers that such an approach will not only help solve soil and water conservation problems but also restore wildlife habitat and promote tourism. The alternative option, discussed in section 3.5 below, is to reduce the level of price support while maintaining agricultural income via payments to farmers in a manner which does not encourage further production.

As well as creating surpluses in some situations, product price support policies can also cause distortions between products and sectors. The proportion of land which is devoted to forestry and the proportion devoted to agricultural production is one example of this. In Sweden, although the policy is under review, agricultural land may not be planted to forest trees without permission, while in Germany, forest lands may not be converted to agricultural lands. Concern is being expressed in several countries about the impact which all these policies have on wildlife habitat, landscape amenity and regional employment opportunities. The United States, for example, is encouraging the planting of trees, plants and wildlife habitat on its conservation reserve lands. The issue of the merits of extensive plantations versus native timber forests is also receiving considerable attention in several countries. Portugal is particularly concerned about the impact which forest monocultures have on the environment and is seeking ways to promote species diversity in reafforestation.

For many years the less productive northern part of Sweden has been compensated for its higher agricultural production costs by the payment of a differential price for products produced in this region. This has been done partially to maintain production and support regional employment. It has also been used to prevent parts of the landscape in this region from reverting from agriculture to forestry. In Sweden, differential prices are also used to regulate agricultural production. In other marginal areas within Europe there have been calls for similar differential price arrangements to maintain landscapes in less favoured mountain areas but, as yet, none have been introduced. Payments to farmers in less favoured areas which are not linked to product prices, however, are common in many European Community countries.

In the past Japanese agricultural policy was implemented on a sub-sector basis but, since 1982 a new "Integrated Programme for Agricultural Production Improvement" has been implemented. Under this programme each municipality prepares a comprehensive production plan with a view to making the best use of its soils and harmonizing agricultural activities with environmental protection. Government subsidies are then paid to the municipality to enable the plan to be implemented.

Policies directed at the capital structure of agriculture

Development controls

Most countries in their replies have used a narrow definition of "capital structure". A broader definition would have included consideration of the impact of transportation facilities such as railways on agricultural development and the environment and also such factors as the effects of land tenure provisions on agriculture and the environment.

Measures taken to control development can be roughly divided into "development controls" and "incentives". The French report, while pointing out that most measures directed at the incorporation of environmental objectives into agricultural policy in France are of a research, extension and incentive nature, suggests that in certain circumstances statutory land-use controls are effective policy instruments for ensuring that agricultural development is compatible with environmental policies.

Among controls "planning permission" is straightforward and used by some countries. For example, the United Kingdom has Town and Country Planning Acts which, although they exempt most agricultural development from control, require farmers to obtain permission to build farm houses. Consideration is now being given to strengthening these laws to facilitate control over the location of new livestock buildings close to residential or similar property. Thus, planning controls can effectively prevent the scattered building of houses in agricultural areas. Zoning is also used in several countries to control agricultural land-use by defining permitted uses and excluding urban and industrial development from certain areas. Legislation is also used to protect farmland from urban development in countries like Portugal where, since 1982, a special permit has been required to develop land for urban or industrial purposes. In Belgium, large scale intensive crop growing and the application of agricultural chemicals are also either restricted or prohibited within certain agricultural zones.

In national parks in Europe agricultural development is now usually subject to strict planning control. In France, certain developments such as large-scale livestock units, require impact studies before they are authorised. Greece also has regulations for physical planning. Denmark has a system of plans prepared by 12 regional councils and approved by the Environment Minister, that cover such matters as water abstraction, preservation, raw materials, waste and water quality. There are also plans to protect prime agricultural land from urban development in many countries. Japan designates agricultural promotion areas and within these, the prefectual governor may prevent landscape changes and introduce other land use restrictions to preserve and improve the environmental amenity of a region.

Some countries have controls on land ownership designed to favour "family farms". In Germany, for example, farmers and their descendants usually have the first right to purchase farm land. Such controls are by no means always adopted for environmental protection reasons. More often these controls are introduced for political motives of one kind or another, for example when the units are regarded as more socially desirable or politically stable. Nevertheless, "family farms" are generally regarded as being more favourable for the environment than "commercial" holdings. As the family regards the land holding as its home and as there is normally more low cost labour available, time is taken to safeguard the beauty of the landscape. There is also less motivation to introduce labour saving practices such as large-scale mechanisation, excessive dependence on chemical pest and disease control and hedge removal. The units are smaller, so there is less tendency to introduce large-scale intensive livestock enterprises. Norway limits livestock numbers on farms as one means of implementing such a policy.

Finland also favours family farms and protects them by requiring farmers to reside on or near the land they purchase. This policy, however, is also considered in parallel with a recognised need to increase farm size and reduce the number of sub-economic holdings. No concessional loans are given for the acquisition of holdings below a prescribed minimum size. Permission for companies or institutional owners to buy areas of land in excess of 2 hectares is usually refused.

117

Incentives

Tax concessions are among the principal incentives used to attain environmental objectives. Land tax policies vary from country to country and appear to influence the allocation, development and use of agricultural land. High land taxes in countries like Australia tend to force potential agricultural land into productive use while low taxes in other countries encourage farmers to maintain woodlots and other areas of undeveloped land. Land used for forestry is often taxed favourably, for example in the United Kingdom. Undesirable land use may lead to tax penalties: in France land clearing is taxed and inheritance taxes are reduced on forest land. In Canada on the other hand, the tax system is generally neutral to conservation.

In Australia, tax allowances and advantages for land clearance have been abolished but have been introduced for tree planting. Soil conservation also draws tax allowances as does land that is left uncleared under a Heritage Agreement. Conversely, and coincidentally, the introduction of capital gains taxes could have beneficial effects for the environment. In many areas of the United States, there are tax concessions at both state and local level to preserve open land.

In the case of peri-urban land several countries have introduced schemes to promote the retention of farmland by discouraging its conversion to urban and industrial uses. These schemes are often implemented at a local or regional level. In the United States, for example, most states allow farmland property to be assessed at its current agricultural use value rather than its market value; some states permit developers to increase the density of development by purchasing development rights from farmers; and a few states have introduced legislation which permits the State to purchase development rights from farmers by paying them the difference between the market value and the agricultural use value of the land.

Some countries have found that it is difficult to target tax policy so that it has the desired effect in all situations. Consequently there is a tendency in several countries to adopt a neutral tax approach to considerations which involve agriculture and the environment. New Zealand is phasing out taxation allowances that favour land clearance. The reduction of allowances will also make it less attractive to substitute capital for labour and thus reduce rural depopulation.

Grants or concessional loans are made in a number of countries for works and practices that favour the environment. In Japan, for example, subsidies are made available to farmers for the construction of cooperative facilities for processing animal manure. Interest free loans are also provided to enable farmers to purchase machines which, by inserting fertilizer next to rice seedlings during transplantation, reduce water pollution.

Another approach is to make various income support and capital development grants conditional on the protection or improvement of the environment. Sometimes these are only made available for a transitionary period. For example in Sweden, when regulations requiring improved manure storage facilities were introduced, grants for upgrading storage facilities were made available over a transitionary period. Norway makes similar grants available to farmers so that they can upgrade storage facilities but has not announced when these grants will be withdrawn.

Recently the European Community modified its socio-structural policy to provide measures for the protection of the environment. Community aids are foreseen in particular for investments related to the protection and improvement of the environment on agricultural holdings. In the same line, Community compensatory allowances can be given to farmers operating in small areas affected by specific handicaps and in which farming must be continued in order to ensure the conservation of the environment and the countryside. National aids are authorised to farmers who contribute towards the introduction or continued use of agricultural production practices compatible with the requirements for conserving natural habitats and ensuring an adequate income for farmers. The United Kingdom has introduced such national aids in a number of areas and intends to expand this programme in 1988. The Community supplements these national aids with its own financial support.

The United Kingdom also has a protection system for sites of special scientific interest or of landscape value, under which management agreements are made to safeguard the particular environmental interest. An example is the Broads Scheme, affecting a sensitive valuable wetlands zone that is farmed. In return for an annual payment, farmers undertake not to use certain practices and to follow an agreed system of management. Similar agreements can be made in areas where the environment is at risk throughout the European Community.

In the United Kingdom, capital grant schemes for the improvement of agriculture include grants for works of benefit to the environment such as hedge planting, building traditional walls and, in some places, planting shelter belts. Grants for hedge removal to make larger, more efficient fields, have been withdrawn. Proposed projects under these schemes in National Parks, nature reserves, the Broads area and sites of special scientific interest have first to be approved by the relevant authorities. Grants are made for works that enhance the environment and to help farmers achieve a better balance between environment and agriculture.

In Switzerland, interest free loans made for agricultural improvements and land consolidation now take much more account of environmental considerations. In Belgium, grants are made so that egg and poultry production can be made more environmentally acceptable, provided that production is not increased. European Community policy now rules out grants or concessions to poultry farmers for production increases. Similarly, community members may not pay grants to farmers who run more than 550 pigs. In Finland, to qualify for a loan, buildings must be constructed to blend in with the landscape.

To reduce the intensity of production and enhance environmental values a wide variety of production limitations is also being introduced. One example of this is the requirement in Finland that permission be obtained from the National Agricultural Board to establish a holding above 60 beef cattle, 200 eight-week old pigs, 1,000 laying hens or 30,000 poultry. Along similar lines but with the additional aim of protecting rural industries an act was passed in Austria in 1983 to require farmers to obtain permission to keep more than 400 fattening pigs; 50 breeding sows, 130 fattening calves; 10,000 laying hens; 22,000 fattening hens; 22,000 pullets; or 12,000 turkeys. When more than one of these types of animal is kept a proportional adjustment is made to these limits. In Japan, in regions where odour and other environmental problems associated with intensive animal husbandry exist, assistance is given to farmers to encourage them to relocate to areas where their facilities will have less impact.

Another important development in recent years has been the adjustment of land consolidation schemes to ensure that they include the full consideration of environmental considerations. In Belgium, for example, basic ecological studies now play a part in the development of land consolidation projects. Similarly, in 1967 German land consolidation procedures were modified to enhance landscape amenity and nature conservation values.

Finally, many countries put emphasis on research and technical assistance in the area of capital structure. In France, a high level officer has been appointed to oversee the important problem of land drainage in ecologically fragile wetland areas. Local sensitisation has led to the appointment of landscape architects and counsellors. Regional and local development frameworks have been produced. Increased decentralisation has aided this process. Local groupings of interested parties such as farm organisations, industrialists and those concerned with environmental issues give a community-wide input to plans. In the United Kingdom advice on environmental questions is given both by the Agricultural Development Advisory Service and increasingly by Farm and Wildlife Advisory Group Advisors. The Countryside Commission and the Nature Conservancy Council have responsibility to advise and promote environmental concern in addition to their grant-giving role.

Overall in this section it can be said that a wide range of measures, both coercive and incentive, are used to influence the capital structure of agriculture and that, increasingly, more and more consideration is being given to environmental considerations in implementing them. Many environmentally-orientated measures have been adapted from farm support measures or wider political measures.

Policies directed at the positive and negative environmental effects of agriculture

Several governments, notably France, Germany and Switzerland, stressed in their papers the potential for positive impact on the environment of agricultural activity, including the development and protection of the landscape for recreation, and the protection of sensitive biotopes.

Switzerland recognises that the landscape, as valued by contemporary society, is largely a man-made creation. The modern tendency to uniformity and the abandonment of difficult areas are both now seen as undesirable trends. The maintenance of the traditional customs and infrastructure of rural life in the Alps and the Jura has both social value and economic value for tourism. Erosion, avalanches and flooding are all to some extent controlled by agriculture. In Sweden, while the emphasis is more on statutory controls than in many countries, agricultural education has an increasing environmental content.

The French Ministries of Agriculture and of the Environment have issued a joint declaration covering such matters as the management of problem areas like mountains and wetlands, the management of genetic resources, the protection of water and soil and training. Canada has developed a federal strategy to deal with soil and water degradation. Jurisdiction and programming are shared with the provinces and discussions on a National Strategy are continuing.

In Finland the environmental impact of agriculture is generally regarded as beneficial. Publicity for this positive relationship is generated by competitions and prizes, for example for the best landscape or best farmyard. New Zealand has farm water and soil conservation plans and catchment management plans.

Public education campaigns

Most countries use various public relations approaches to persuade farmers of their interest and duty to use or adopt farming practices that are more favourable to the environment. In the United Kingdom, the adoption of environmentally appropriate practices by farmers relies on incentives, encouragement and persuasion. Britain's public education strategy involves inputs from a wide variety of organisations such as the Ministry of Agriculture, the Countryside Commission and the Nature Conservancy Council. Non-government organisations also play a significant role. The approaches taken include the appointment of county conservation advisers, the formation of farm and wildlife advisory groups, the distribution of publications, the provision of advice on conservation practices and the organisation of events with specific environmental themes. Grants are made to various non-governmental bodies. Agricultural advisory officers are encouraged to act as secretaries or otherwise assist these bodies. Germany also uses the extension service to advance environmental protection.

In Canada and the United States, an annual Environment Week, a National Soil Conservation Week and a National Forestry Week are held. Australia also has public education campaigns, including a National Tree Program, a Rainforest Conservation Program, a National Soil Conservation Program and a National Conservation Strategy.

Regulations and incentives

Some countries take a highly regulatory approach to controlling the negative impacts of agriculture on the environment, with extensive legislation and regulations for protecting the environment. Many of these regulations restrict the use of environmentally active inputs such as pesticides, and have been considered in detail above. Among countries that favour the statutory approach, Sweden perhaps goes furthest. In this country there is an Environment Protection Act, an Act on Hazardous Products, a Nature Conservancy Act, a Water Act and a Hunting Act. Furthermore, there have been additions to the Agricultural Land Management Act that have nature conservancy elements. Discharge of liquid manure and silage liquid is controlled according to plans agreed and checked by environment protection staff. As a result of collaboration with farmers' organisations, this system is reported to work well. The further development of drainage systems is strictly controlled although existing systems can be maintained. Working out the operation of these controls has improved mutual understanding between the various parties.

There are many other controls on farmers' actions. For example, headlands, stone walls and avenues must not be destroyed within French National Parks. Ponds must not be drained or filled with soil. Fertilizer may not be applied to hitherto unfertilised meadows in some protected areas of the United Kingdom. Some pesticide use is restricted to agricultural land.

Other examples of a broad legislative approach are the Clean Water Act in Canada and the Resources Conservation Act in the United States, planning controls and water pollution regulations in the United Kingdom, standards for water bodies in Australia's State of Victoria and clean water legislation in other Australian states.

In Australian, Canadian and American rangelands, restrictions on land use to achieve environmental objectives are often implemented through the tenure system. Some pastoral leases in Australia and grazing leases in the United States, for example, contain provisions which protect flora and fauna and require forage to be left for native animals.

In general, while the Polluter-Pays Principle has been accepted by OECD countries, there are some reservations as to its application to agriculture. For instance the United Kingdom reduces point sources of pollution by providing grants for the purchase of some pollution avoidance equipment. Denmark accepts the Polluter-Pays Principle, but was granted a transition period for implementation agreed by the European Community. From early in 1988, following a two-year period of review and a moratorium on development, The Netherlands is now intending to enforce the Polluter-Pays Principle. While several governments believe that agricultural units should bear the greater part of the costs of pollution control, some financial help from governments may be needed in order to avoid distortions, in accordance with the Polluter-Pays Principle. Finland also supports the Principle but does subsidise the costs of liquid manure storage. Australia is moving towards full cost pricing of irrigation water as a method to control salinisation, but has not yet begun to price water so as to recover the full cost of downstream salinity control measures and water filtration costs.

In fact, a majority of countries pragmatically provide subsidy payments in one form or another to help and encourage agriculture to protect the environment and reduce pollution. The United Kingdom makes payments under management agreements in designated sites of special scientific interest and also in environmentally sensitive areas. Norway makes finance available at favourable terms and provides tax concessions for investments that protect the environment. Denmark subsidises the upgrading of manure storage facilities, but an earlier proposal to tax chemical fertilisers to pay for this has been rejected. In Germany some Laender make payments on the basis of difficulties and losses to farmers and subidise the use of special farming methods to protect flora and fauna. Easements, lease back arrangements, tax rebates and out-right purchase are among measures being considered and used in many other countries.

Australia has a low level of agricultural subsidies by OECD standards, so the need for subsidies to promote the adoption of environmentally favourable agricultural practices is also low. New Zealand reports that it has had similar experience in this regard. In contrast, Switzerland has a clear policy of subsidising certain desirable agricultural activities. Headage payments are made on livestock in mountain areas. There are payments for the satisfactory cultivation of slopes steeper than 18 per cent, for maintenance of traditional pastures and summer pastures. There is legislation to allow third parties to cultivate abandoned land. Livestock housing and manure storage are also supported in Switzerland. Even in Sweden there are incentives to abandon milk production. Reduced cattle numbers have allowed forest to take over meadows and there are subsidies to maintain them.

Where there are substantial potential benefits to society from a changed agricultural practice but which, from an agricultural viewpoint is unlikely to increase profits, some countries are subsidising or paying farmers to adopt alternative practices. Generally, the types of payments made are either a direct grant to the farmer, a payment per hectare or a payment per head of livestock in a manner which encourages the adoption of environmentally favourable practives. Specific examples of these are:

-- the payment of farmers in the Tauern National Park situated in Salzburg Province for fulfilling tasks involving the protection of nature and landscape;

-- the payment in Canada of incentives to maintain waterfowl and other wildlife habitats;

-- land tax concessions in Australia for farmers who enter into a heritage agreement to preserve a special landscape, wildlife habitat, etc;

-- the European Community's contribution towards the introduction or continued use of practices which conserve natural habitats, natural resources, the landscape, flora and fauna; and

-- compensatory allowances, under Directive 75/268/EEC, for the conservation of the countryside.

These practices would suggest that in many countries when society wishes farmers to improve certain aspects of the landscape or undertake conservation efforts, it could consider paying farmers for the cost of providing these services. In Austria, Switzerland and also in France, it is considered important to retain cultural particularities and traditional landscapes which are the outcomes of centuries of historical development. It is also seen as being essential to safeguard leisure and tourist areas. In the Alps, for example, mountain grazing and hay cutting plays an important role in reducing the risk of avalanches. Consequently, in Austria the federal and the laender governments provide grants to mountain farmers to tend and preserve mountain landscapes.

A number of non-governmental organisations have also begun to pay and encourage farmers to adopt or, alternatively, agree not to adopt certain practices. For example, in Austria a major non-government organisation has leased extensive pasture areas and pioneered the solution of nature protection problems associated with agriculture. Similarly, in the United Kingdom, the National Trust has raised £8.5 million and bought 465 miles of coastline. Wherever possible the land is acquired and then resold with suitable covenants to protect environmental interests. Similar arrangements are also becoming more common in the United States and Canada.

Research and Monitoring

Research and monitoring form a large part of the effort of many countries and provide an important opportunity for countries to influence gradually the direction of agricultural policy. In many countries it is now common practice to appoint people with environmental interests to boards and committees allocating funds for research. For example, Dutch councils for agricultural research and, also, for environmental and nature research are composed of representatives from government, research and other interest groups.

In the United Kingdom, the Countryside Commission, Nature Conservancy Council, Development Council and the Sports Council all support research. The Department of the Environment does wider in-house research and also sponsors research. The Ministry of Agriculture, Fisheries and Food has a system of consumer groups advising its Chief Scientist, which leads to the inclusion of conservation and environment interests in the research planning process. The effect of nitrogen on biological diversity in grassland is a subject being studied in the United Kingdom, and an integrated rural development project involving co-operation between agencies and a trial alternative grant scheme is underway in the Peak District National Park.

Norway has found data on non-point sources of pollution, especially phosphorous compounds, is inadequate and, as a consequence, is financing a nationwide action programme to develop models which, using cost-benefit studies, identify cost effective methods of reducing non-point sources of pollution. Research on the relationship between soil, water, plants and the environment is also being conducted by agricultural research stations, by universities and colleges and by the State Pollution Control Authority.

In Denmark most agricultural research is carried out by the Ministry of Agriculture. Its main aims are improving agriculture's competitive position, increasing employment and earning foreign exchange, but protection of the environment is increasingly being considered. Denmark also has launched a special action programme on the identification of alternative production methods.

France puts particular emphasis on research, on disseminating the results and on informing and training farmers. There is co-operation, for example, between Ministries on research on nitrate pollution. Research subjects include the impact of agriculture on the environment, the improvement of agricultural production methods, the development of new varieties, the better use of nitrates, and the development of improved agricultural techniques for less-favoured areas. The Environment Ministry is now involved

124

in the management of the main research bodies such as the National Agricultural Research Institute (INRA) and the Agricultural, Rural, Water and Forests Engineering Centre (CEMAGREF), and can also commission research. In its turn the Environment Ministry has set up research committees on which the Ministry of Agriculture is represented.

Canada has a programme of research, demonstration and extension of improved agronomic practices, and monitors water and land use. Australia has recently recognised the importance of non-point sources and is carrying out surveys and monitoring in which both government and industry are cooperating. Research at the federal level is done by the Commonwealth Scientific and Industrial Research Organisation, together with state level institutions. Finance to promote this research is available from rural industry research funds and a levy on agricultural products.

Japan stresses the contribution that many agricultural practices make in purifying water, recycling nutrients, preventing floods and preventing soil erosion. Consequently, its research activities include the development of methods to retain soil structure and promote perculation as well as the use of oxygen produced by green algae in agricultural drainage waters to reduce water pollution.

Switzerland has given a new orientation to its research policy, putting more emphasis on quality and stressing the national use of fertilisers and organic farming. Subjects investigated include the water cycle, socio-economic and ecological aspects of farming in the mountains, sewage disposal, use of wood, air pollution and soils. Belgium and The Netherlands also are encouraging research on environmental questions and the dissemination of research results.

Greece is monitoring non-point pollution by watersheds, particularly drinking water from rivers and lakes. The Ministry of Agriculture is financing special research, for example into olive oil plant waste. Finland is monitoring the effects of agriculture and forestry, particularly the pollution of waterways from spot-sources. There is a proposal for cultivation-free buffer zones along water courses.

New Zealand and the United States are monitoring erosion and have land resource inventories. In pursuit of similar information, Portugal is now planning to increase the number of sediment transport measuring stations installed in areas where erosion is a problem.

Another set of controls is the protection of areas of special environmental value. In the United Kingdom these include national parks, areas of outstanding natural beauty and sites of special scientific interest. In Germany nature and water protection areas are designated by the laender. The United States has wilderness areas and parks at both the national and state level, while in Canada most park areas are under federal jurisdiction. Spain has significant wildlife reserve areas, particularly wet lands. The degree to which agriculture and other development is allowed in such protected areas varies from country to country and according to the type of land being protected. The essential common point is that in areas where it is desired to protect natural resources of special interest, this is usually done by either statuatory control or, in some countries, the outright purchase of the land in question. In such cases, agricultural and other interests are normally subordinated. One notable difference amongst OECD countries is that some countries exclude agriculture from national parks and other protected zones, while others consider agriculture to be an integral part of them. For example, in France there are now 23 national parks which seek to preserve agriculture as an integral part of the ecosystem. Many European flora and fauna are dependent on agriculture for survival. In the United Kingdom, for example, research has revealed that the scarce large blue butterfly, which is now extinct in The Netherlands, depends for its survival on grazed and trampled pastures which contain thyme or marjoram.

In France, assistance is given to municipalities, rather than to individuals. An example is drainage schemes, which are carefully studied for their environmental impact. Grants are made for environmental enhancement within the framework of the European Community, such as buildings for livestock in mountain areas. There are also grants for agriculture in "Less Favoured Areas" one of the main purposes of which is to protect the environment by maintaining agricultural activity.

4.   SUMMARY

This report of information prepared by countries on approaches they are taking to integrate agricultural and environmental policies summarises the first part of the project being undertaken by the ad hoc Group on Agriculture and Environment for the OECD's Environment Committee and the Committee for Agriculture. Other work on specific policy areas, which is underway in the second part of the project, will form the substance of following reports. The results of the work on these policy areas will then be combined with the information described in this Report and evaluated in the ad hoc Group's Final Report on the Integration of Agricultural and Environmental Policies.

Perhaps the most consistent and notable aspect appearing throughout the country information papers was the awareness, which has grown in the last decade, that the agricultural sector can be both a significant polluter and a major protector of the environment. One reason for this is that in many countries there has been an essentially industrial and urban focus on environmental policy dealing, over the past three decades, with major environmental problems in the cities where the bulk of affected populations and main pollution sources are concentrated. Another reason was that some of the long-term effects of pollution arising from agricultural activities, such as the pollution of ground water sources, have often only recently been detected. A third reason is that agriculture has expanded rapidly over the last decade. This has heightened awareness of issues facing the agricultural sector on the one hand, and on the other, environmental issues stemming in part from the increased production and intensification of agriculture.

Institutional arrangements

The overall impression gained from this initial review of the integration of environmental and agricultural policies in OECD countries is that there are a wide variety of administrative and institutional arrangements forming the framework within which policies are formulated and implemented.

Most often, the basic model comprises both a Ministry for Agriculture and a Ministry for Environment surrounded by several other departments interested in, and affected by related environmental, agricultural and natural resource issues. The situation is, however, continually evolving: several countries have only just created a Department of Environment. Despite many different models, administrative frameworks can be, and have been found to facilitate the better integration of agricultural and environmental policies. But such arrangements do not exist in all countries.

Innovative administrative structures and approaches are needed and are being found to facilitate the integration of agricultural and environmental policies. One innovative example is the involvement of three Dutch Ministries in the development and implementation of an Indicative Multi-Year Program for Environmental Management which identifies central themes and develops "effect-orientated" and "source-orientated" policies. Other examples are the creation of administrative frameworks which facilitate the public debate of policies and the creation of quasi-autonomous statuatory bodies with responsibilities for conservation.

Increasingly, the benefits from inter-departmental consultation before and during the policy formulation process are being recognised. The benefits of prior consultation with, and involvement by farmers and other private organisations have also been demonstrated.

The content of agricultural and environmental policies is rapidly changing. Many governments and the European Community have now incorporated a clear mandate for preserving, maintaining and enhancing the environment among their official objectives for the agriculture sector.

Governments are also recognising the value of obtaining estimates of the off-farm cost of agricultural pollution and of assessing the net social benefits of alternative policy options. Cost-benefit and cost-effectiveness studies are increasingly being used as a mechanism to achieve the better integration of agricultural and environmental policies during policy development.

Instruments for managing inputs

Agricultural inputs, in several countries, are a significant source of environmental problems. Standards are used in all countries to control specific problems, especially those associated with the use of pesticides. There is considerable divergence as to input pricing policies, particularly regarding the use of quotas, charges, levies and taxes as a mechanism to reduce the off-site impact of these inputs. Several countries are also using such charges on inputs simultaneously to reduce agricultural surpluses and improve the environment. There is, however, a general consensus that farmers should be encouraged to avoid the over-use of inputs.

In several countries, extension and advisory officers are finding that some farmers lack detailed knowledge about the responsiveness of crops to fertilizers. In some countries pollution of certain waters has been recognised to be a consequence of over-fertilization, especially on farms using both animal and mineral fertilizers. This problem is being addressed in some of these countries by the development of more effective technical advisory services, especially since a more rational use of fertilizer can increase net farm income. Others are developing integrated strategies which involve a mixture of taxation and advisory policies. There have also been significant advances in the development of integrated pest management systems and the development of joint integrated manure and mineral fertilizer plans.

Farmer acceptance is essential to the success of advisory programmes and, increasingly, countries are designing extension programmes to make farmers more aware of the contribution they can make to the environment. The benefits of programmes based on voluntary participation are stressed by many countries. In many situations, especially when point sources of pollution can be identified and where the intensity of agriculture is increasing and off-site impacts are significant, regulations are being used to reduce some of the potentially adverse impacts of agriculture. During the transition period which follows the introduction of new regulations, it is common for countries to offer economic incentives to encourage farmers to make changes which are desirable from an environmental point of view but do not yield immediate economic benefits to farmers.

Regulations are also being introduced to ensure that the further intensification of agriculture does not exceed the environmental capacity of a region. Several countries, in the face of simultaneous surpluses and undesirable environmental impacts, especially on the rural landscape, are now searching for ways to reduce the intensity of agriculture and to encourage environmentally favourable agricultural practices.

Instruments associated with product prices

There is a wide divergence in opinion among OECD countries about the effects on the environment of production incentive policies. The information presented suggests that, in the past, most support policies were considered without consideration of their environmental consequences. Generally, the principle reasons for their introduction were a) the maintenance of adequate farm income and b) the goal of self-sufficiency.

Some countries consider that support measures increase both the level and intensity of agriculture and, also, lead to greater use of inputs with, as a consequence, an attendant risk of negative effects on the environment. Others point out that there are also cases where these policies have enabled farmers to continue agricultural practices which are essential to the maintenance of the environment.

Support measures have tended to encourage agriculture in marginal areas which would otherwise be left unfarmed. This is by no means always a negative result. Often it is felt that continued farming of mountainous areas contributes to their environmental value. On the other hand support has also led to the cultivation of areas such as wet lands which were highly valued.

In the last few decades a significant loss in landscape amenity has been experienced, substantial soil erosion has occurred or significant water pollution has arisen from agricultural sources in a large measure due to increased agricultural activity. There is increasing uneasiness about such developments when faced with growing surpluses.

As a consequence of the above problems and to encourage the adoption of environmentally favourable farming practices, management agreements and other alternative forms of income support are being considered and in some cases used to maintain environmentally desirable agricultural practices. Payments per hectare and for the adoption of specific practices are becoming more common. It may be desirable to maintain such arrangements in less favoured and environmentally sensitive areas if the prices received by farmers for their products are reduced.

Other approaches to better integration

Some countries have used taxation policy to promote the integration of environmental and agricultural considerations but other countries consider that taxation policy, already heavily burdened by other policy considerations, should be neutral to environmental objectives. Land-use regulations may be more efficient than tax policy in limiting agricultural development and ensuring that the structure of agriculture contributes to the environment.

A recent development has been the renewed interest in land set-aside policies coupled with a concern for "abandoned land". In fact, such land is rarely abandoned by its owners. Rather, its use changes from cropping to grazing or forestry, although occasionally it is simply left idle. Forest fires, the loss of scenic amenities and wildlife habitats can all be the result of such changes. Often when the holding costs are low such land is inadequately managed. A number of countries are now considering the possibility of using such land for nature conservation and protection purposes. The interest in set-aside programmes has been stimulated in large measure by the emergence of agricultural surpluses and a desire to reduce them. Building on past experience with land set-aside programmes, one country has introduced an innovative concept of cross-compliance. Cross-compliance limits eligibility for government support programmes to farmers who meet prescribed environmental objectives.

Many countries are devoting considerable effort to identifying the most appropriate way to distribute the cost of agricultural pollution between farmers and society. There is consensus that in the case of point sources of pollution, the "Polluter-Pays Principle" should apply to agriculture, with appropriate transition periods. Where non-point sources of pollution are concerned, most countries are attempting to divide the costs in such cases among society, those affected and the farming sector. But some countries are proposing that most of the costs of pollution from the agricultural sector should be internalised. Suggestions for nitrate taxes and quotas, for example, are consistent with this interpretation.

Information from countries reflects no consensus on how to apply the Polluter-Pays Principle to non-point sources of pollution. At times the approach seems to reflect a pragmatic effort to remove or overcome a pollution problem as quickly as possible. Countries are united, however, in recognising the importance of maintaining a viable farm sector and ensuring that measures taken by farmers to meet environmental standards are cost effective.

It is also recognised that farmers can and do make a major contribution to the maintenance of agricultural landscapes and environmental quality. This, amongst other things, also contributes to the development of recreation and tourism. Consequently countries are entering into management agreements with farmers to manage landscapes in a manner which enhances their quality for recreation, tourism, wildlife and other amenity values.

There is a consensus that, given the long time lag for environmental benefits to accrue from changes in agricultural practices, there is an urgent need to continue to search for more effective ways to better integrate agricultural and environmental policies. Ideally, policies should always reflect long-term environmental and agricultural goals and should avoid foreclosing opportunities to achieve them in the pursuit of short-term objectives.

Finally, the impression emerging from most of the country information papers is that public intervention should generally not take the form of long-term subsidies of the kind that create environmental damage, or hinder environmental improvement.

ANNEX 2

SELECTED POLICY ISSUES RELATING TO
AGRICULTURE AND ENVIRONMENT

Chapter 1

STUDY FRAMEWORK

INTRODUCTION

Growing awareness and concern about the linkages between agricultural policies and practices and specific environmental problems, some of them with serious long-term implications, have focused increasing attention on the interface between agriculture and environment. This has led to a recognition of the need to better integrate agricultural policies with policies which seek to protect, preserve and enhance the environment. This recognition led to the OECD project on "Integration of Environmental Policies with Agricultural Policies", under the aegis of the Organisation's Committee for Agriculture and Environment Committee.

The agriculture/environment integration issue can be succinctly stated as follows: to what extent do macro-agricultural policies take into account their effect on environmental quality, and to what extent do environmental policies take into account their impact on agricultural output, income and prices. Defining the problem in these terms reflects the growing recognition in both domains of the close linkages and the potential for mutually reinforcing action by agricultural and environmental policy makers. Thus, integration is not a question of bending agricultural policy to meet exclusively ecological goals, nor of shaping environmental policy to suit the needs of agriculture. Rather, policy, by making appropriate trade-offs between the interests of both sectors, can contribute to rational, efficient and sustainable development in both an economic and ecological sense.

An overview of integration of agricultural policies with environmental policies in OECD member countries, conducted as the first phase of the OECD agriculture and environment project, confirmed that the principal areas of concern leading governments to re-examine the interrelationship between agricultural and environmental policies tend to fall into five broad categories:

-- the problems of pollution and environmental degradation associated with intensive crop production;

-- the post-war trend, especially in Europe, towards intensive animal husbandry, creating its own particular environmental problems associated with animal wastes;

132

-- the downstream effects and lost production from soil erosion caused by extensive agricultural practices associated with dryland agriculture;

-- the changing demands and practices that have significantly altered the character of the landscape, leading planners to seeking ways to use agriculture to save dwindling natural habitat and wilderness areas for the benefit of present and future generations; and

-- the impact on agriculture of pollution from other sources and non-agricultural activities, notably in the form of acid precipitation, water pollution and sewage sludge.

Methodology

In the second phase of the OECD project, these five categories of environmental issues associated with particular agricultural policies and practices were selected for in-depth examination. Consultants were commissioned to carry out detailed country studies designed to illustrate the major problems and policy areas at the agriculture and environment interface, based on the latest information on evolving policies and programmes in OECD member countries.

The subject areas and country studies were organised as follows:

   i) Intensive crop production/agricultural chemicals interface
       (Federal Republic of Germany, Sweden, United States);

  ii) Intensive animal husbandry/animal manure interface
       (France, Netherlands);

 iii) Dryland agriculture/soil conservation and erosion interface
       (Portugal, United States);

  iv) Changing landscape and land use patterns/rural quality interface;
       (Austria, United Kingdom); and

   v) The impact on agriculture of pollution from other sources (a review
       of the literature and state of knowledge in this area).

Five of the consultants contributing to the case studies were then asked to prepare the summaries which form the principal content of this report.

Each policy area study began with a description of the environmental, economic and social costs and benefits of the relevant agricultural policies and practices. Within each of the policy areas, careful attention was given to a set of policy issues such as the effects of subsidies on inputs, the need to improve advisory services and the impact of production support measures on landscape amenity. The results of this examination were then compared with policy alternatives and practices which could result in a better integration of agricultural and environmental considerations.

Each study involved the following four steps:

i) the identification of the benefits and impacts which arise from different agricultural practices within the policy area;

ii) the identification of the socio-economic, administrative and physical conditions which cause these impacts or which are necessary to derive these benefits;

iii) an evaluation of the effectiveness of policy instruments which seek to enhance benefits and mitigate negative impacts; and

iv) the identification of the necessary conditions for and approaches to the successful implementation of policies which achieve better integration.

The principal problems and issues exmined in the case studies

Intensive Crop Production/agricultural chemicals interface

The focus of this interface is on the effects on the quality of the environment of the fertilizers, both organic and inorganic, and pesticides used in intensive crop production. The meaning of the term "intensive" varies from country to country, but generally such agricultural systems are characterised by either irrigation and/or high rainfall and the observation that, for at least most of the year, water availability does not limit plant growth. Intensive crops include almost all horticultural and agronomic crops, such as sugar cane, sugar beet, cotton, potatoes, maize, wheat, barley and oats.

The principal environmental problems considered in this policy area study were:

-- the effects of residual fertilizers and pesticides on human health including those associated with spray drift from pesticides during application, and residues in food and water;

-- the effects of residual fertilizers (phosphates and nitrates) on the condition of aquatic environments (eutrophication); and

-- the unintended effects of pesticides on species other than those to which they are applied.

The principal issues selected for close examination were:

-- the relationship between product price support policies and the extent of fertilizer and pesticide use;

-- the impact and effectiveness of policy instruments which seek to reduce the quantity of fertilizers and pesticides used including charges, taxes and subsidies on inputs, quotas on inputs and production standards for fertilizer and pesticide manufacture;

-- the need for and effectiveness of regulations to prohibit the use of certain chemicals and adoption of certain application practices;

-- the effectiveness of advisory services in making agricultural practices more favourable to the environment;

-- the contribution of agricultural research policies to environmental policy; and

-- the identification of "non-point" sources of pollutants.

Intensive animal husbandry/animal manure interface

The focus in this policy area study was on the adverse effects that animal manure, if not properly managed, can have on the environment. These affects are most noticeable in connection with intensive animal husbandry, notably for pigs, poultry and dairy cattle.

The most significant related environmental problems include nitrate pollution of ground and surface water supplies, the effects of ammonia emissions on surrounding vegetation and their contribution to acid rain, and the accumulation of heavy metals in soils. At a local level, the emission of unpleasant odours is an environmental nuisance problem.

The principal issues selected for close examination in this policy area study were:

-- the effects of price support and tariff policies on the location and development of intensive animal production systems;

-- the impact and effectiveness of policy instruments (charges, taxes, subsidies, quotas and regulations) which seek to decrease environmental pollution from intensive animal husbandry;

-- the application of the Polluter-Pays Principle to non-point sources of pollution; and

-- the need for and effectiveness of regulations to control the quality of prepared animal feedstuffs.

Dryland agriculture/soil conservation and erosion interface

The work in this policy area recognised both the positive and negative effects of agriculture on the environment. In particular, it examined the role for agriculture in contributing to environmental objectives through the maintenance of soil structures and productivity. The off-site benefits from desirable agricultural practices such as those which reduce silt loads in surface waters, etc., were also examined.

Dryland agriculture is characterised by the fact that moisture is often the limiting factor for plant growth. Generally, dryland agricultural systems contain either native or improved pastures grown in association with cereal crops.

The principal issues selected for close examination in this policy area study were:

-- the potential and cost-effectiveness of grants and subsidies in encouraging soil conservation;

-- the effectiveness of regulations which require and/or prohibit certain agricultural practices;

-- the identification of rates of soil loss and the problem of farming activity where rates of erosion are excessive;

-- the potential of community-backed education campaigns to raise awareness and change community values;

-- the effect of capital and income tax policies on conservation; and

-- the co-ordination of administrative actions during policy formation and implementation.

Changing landscape and land-use patterns/rural quality interface

The work in this interface focused both on the positive role of agriculture in maintaining landscapes and land use patterns and also the impacts that changes in agricultural practices can have on the quality of the rural environment. Particular attention was paid to the environmental benefits derived from the role of agriculture in promoting tourism, providing wildlife habitats and maintaining aesthetically appealing landscapes. This included an examination of policies designed to encourage the development of these roles in alpine and mountainous areas. The integration of rural development and agricultural policies was also examined.

The principal issues selected for closer examination in this policy area study were:

-- the effectiveness of policy instruments in encouraging farmers to maintain a positive role in producing environmental benefits for society;

-- the effects of price support policies on landscape diversity and biological diversity;

-- the effectiveness of policy instruments which seek to reduce, redirect and/or stop the rate of change in agricultural landscapes and land use practices including subsidies, financial incentives and integrated programmes to revitalise economically depressed regions; and

-- changing institutional arrangements to achieve the more effective integration of agricultural, regional development and environmental policies.

136

Impact on agriculture of pollution from other sources

In several countries pollution from other sources has led to regional declines in the quality of food production, and in other cases, the quantity of food produced. There is a growing body of literature describing such impacts on agriculture, although most of it is based on the extrapolation of the conclusions from laboratory experiments about the effects of acid rain, etc., on regional and national estimates of crop production.

For this chapter, a review was conducted of the literature and state of knowledge about this policy area in all OECD countries, describing and where possible quantifying the likely impacts on agriculture of pollution from other sources, notably acid rain, ambient ozone concentrations, water pollution from other sources, sewage sludge spread on agricultural land, and rising CO_2 concentrations.

Chapter 2

INTENSIVE CROP PRODUCTION AND THE USE OF AGRICULTURAL CHEMICALS

H. de Haen (1)

STRUCTURAL CHANGE IN AGRICULTURE

This chapter summarises studies of intensive crop production and the use of agricultural chemicals in Germany, Sweden and the United States. Each study contains a review of the national situation followed by the detailed examination of the effects of current agricultural policy and feasible policy options on the environment. Information from other countries is also included.

During the past four decades, agriculture in the Federal Republic of Germany, in the United States as well as in Sweden has gone through a dramatic structural change. The area of agricultural land has decreased. Crop production has become more specialised. Significant areas of marginal land have been also converted to arable land or forest. On arable land, there has been a shift from ley farming to grain production. The use of fertilizer and pesticides has been intensified, thus enabling the realisation of rapidly growing yield potential. Animal husbandry has been expanded and concentrated on fewer regions and farms. There has also been a shift in livestock production from cattle to pigs and poultry. In response to a steady outmigration of human labour, farm sizes have grown and investments have been directed towards heavy machinery.

The main reasons for these structural changes are similar in the three countries studied:

-- technical and biological progress with a substantial growth of yield potentials;

----

1. This summary of three policy area studies was prepared by Professor H. de Haen in consultation with a study group which included the authors of the studies, representatives of the German, Swedish and United States governments and a representative of the Secretariat. The German study was written by H. de Haen, H.F Finck, C. Thoroe and W. Wahmhoff. The Swedish study was prepared by K.I. Kumm. The United States study was written by R.M. Wolcott, S. R. Johnson and C. M. Long.

-- changes in factor costs, in particular a considerable increase in labour costs; and

-- government price support for major agricultural commodities, resulting in stable producer prices, clearly above world market levels;

-- land consolidation programmes in a number of regions, resulting in a loss of hedgerows (especially in Germany), of boundary strips and small biotopes.

Taken together, these factors have created incentives for farmers to reduce the diversity of production patterns, intensify the use of agro-chemicals and maintain agricultural production, even in those marginal areas where agriculture might not otherwise be competitive under world market conditions. As a result, agricultural production and market supplies have grown much more rapidly than domestic demand in the three countries studied.

Environmental Impacts

The major impacts on the environment which have resulted from this structural change were again quite similar in the three countries (and indeed across virtually all OECD member states). The most noticable of these are the reduction of species diversity caused by the diminution and partition of natural biotopes and habitats; the contamination of groundwater by nitrate and pesticides; the compaction and erosion of soil; the eutrophication of surface waters; the increased detection of pesticide residues in food; and air pollution from dust and odours produced by certain agricultural practices. Of course, dimensions and priorities are not the same in the three countries. The impacts are listed here in the perceived order of harmfulness under current German conditions.

Hence, in Germany, disappearance of species and nitrate as well as pesticide pollution of groundwater are currently causing greatest concern, at least in some of the intensive crop production areas, including horticulture and vineyards, as well as in areas of high livestock density. In Sweden, on the other hand, greatest concern is expressed about the pollution of groundwaters and surface waters, including marine waters, mainly by nutrients but also by pesticide residues. At the same time, however, in Sweden, emphasis is given to the positive role which agriculture plays by maintaining an open landscape in areas already dominated by forests.

In the United States study, it is the human health risks which rank as high as decline of species and groundwater contamination among the external effects of agricultural production. The National Academy of Sciences and the Environmental Protection Agency conducted research which resulted in quantitative estimates of cancer incidence attributable to certain ingredients used in pesticides. Erosion and run-off of residues from fertilizer and pesticides are another major environmental hazard mentioned.

In European countries as a whole, high levels of nitrate in drinking water, attributable in large measure, it is believed, to the over application of mineral fertilizer and the growing concentration of animals, are a source of increasing concern. Austria has acted to restrict fertilizer spreading in some water catchment areas. Heavy metals found in fertilizers such as cadmium are of concern in The Netherlands and Denmark and elsewhere, the presence of cadmium and other heavy metals in sewage sludge used as fertilizer has led to restrictions.

Not all of the main environmental problems of the three study countries are dealt with in the detailed presentation of the respective study regions. The German regions were intentionally chosen to illustrate the environmental problems related to intensive crop production. Both regions have little forestry and a low livestock density. Hence, they represent neither the light soil-livestock areas in the Northwest where nitrate leaching is already causing great concern, nor do they typify many of the hilly areas in the center and south of Germany, where the maintainance of open landscape tends to have a higher priority.

Both Swedish study regions were selected because they had been officially declared as being particularly susceptible to pollution. The availability of plant nutrients in the connected lakes and sea has resulted in plankton algae developing in large quantities (eutrophication). In comparison with the rest of Sweden both regions have a comparatively high livestock density of one livestock unit per hectare.

The United States study region is the upper Midwest cornbelt which produces over two thirds of the country's corn and soybean output. While dietary risk is the major national issue related to pesticides, the principal concern in the region itself is groundwater degradation from both fertilizers and pesticides.

Current Policies and Legislation Affecting
Intensive Crop Production and Agro-Chemicals

All three countries have a broad range of legislation directly or indirectly affecting agriculture, whether intensive or not (e.g. conservation, land use, toxic chemicals, air and water pollution, food quality and health legislation). More recently, new legislative, regulatory and fiscal measures have begun to focus more specifically on problems linked to intensive crop production and the associated use of agro-chemicals, in all three countries. One American State is studying a pesticide tax system, the revenues from which could be devoted to managing conservation reserves or to establishing an information clearing house/education programme related to integrated past management.

The Swedish Environmental Protection Act was amended in 1984 to strengthen the protection of areas "especially sensitive to pollution" by creating provisions to control fertilizer use and cultivation practices. Special legislation deals with the spraying of pesticides and spraying from the air, for example, is banned. The introduction of further bans on filling or cleaning sprayers on lakes and waterways have been introduced.

In Germany, Federal waste disposal legislation contains regulations on the spread of sewage sludge and excessive use of manure. Almost all Laender have so-called crop edge programmes which subsidise farmers who leave predetermined crop edges unsprayed. Eight out of eleven Laender offer grassland extensification schemes which subsidise farmers who refrain from spraying pesticides, reduce levels of fertilizer or leave meadows unused during main periods of hatching.

The other aspect of recent agricultural policy evolution with a direct and potentially important impact on intensive agriculture is the search for means to reduce surpluses. Past policies of price support have led to a large surplus of agricultural commodities and the continuation of price guarantees has engendered severe financial problems for the European Community. As of yet there have been no fundamental attempts to phase out price support in favour of other income transfer policies for the farm sector with fewer incentives for surplus production. The Commission of the European Communities has begun to restrict its hitherto unlimited market intervention, but the effective rates of protection on major agricultural commodities continue to be high, and, hence, continue to stimulate intensive production systems, potentially sensitive to pollution.

In this regard the United States is considering approaches which would decouple payments from specific crops, thus moving towards an income maintenance scheme. In the European Community, as well as in Sweden, attempts have been made to reduce market surpluses without abandoning price supports by various forms of limits on production. Quotas, for example, have been introduced on products such as sugar and milk in the European Community.

Measures to curtail excessive production by withdrawal of land may not be the best solution to the problem of surpluses if they are combined with continued price support, thus motivating more intensive production on the remaining farm land. From an environmental perspective there is another concern with land set aside programmes. Examples are given in the German and the Swedish studies showing that the existing programmes either do not utilise or under-utilise potential to combine reduced production with a better environment:

-- Subsidies are paid in Lower Saxony, Germany, for short run fallow although this increases leaching and has no ecological value;

-- German and European Community programmes to pay a premium on the withdrawal of land farm production for more than five years have no provision to ensure that land set aside would be concentrated in areas needed for nature preservation; and

-- Swedish proposals to subsidise cultivation of energy forests and ordinary forests do not contain provision for siting such activities in environmentally suitable areas.

Policies which, in principle, are more likely to combine a reduction of surpluses with an improvement of the environment include levies on excessive use of fertilizers and pesticides. In Germany, a tax on nitrogen mineral fertilizer was proposed by the Council of Environmental Advisors, but implementation has so far not been seriously considered. Finland, Sweden and Austria have taxed fertilizer and pesticides for a number of years. In the case of nitrogen and phosphorous the Swedish levy amounts to 25 per cent.

Empirical estimates of the impact on input level and on nutrient leaching are not yet available. The Swedish and the German studies, however, present results of production function analyses and farm programming models for the study regions which tend to indicate that charges would have to be very high in order to effectively reduce surplus production and bring environmental relief. Moreover, it should be mentioned that in Sweden, the revenues from such taxes (similar to the Austrian system) are used for extension and research on improved production systems (20 per cent) as well as for the subsidisation of exports (80 per cent).

Quite promising in terms of environmental benefits, but so far with negligible impacts on overall production levels, are various extensification programmes in Germany. Both German case study areas provide examples of bilateral agreements between farmers and regional authorities, which provide financial compensation for certain extensification measures, such as unsprayed crop edges, postponed operations on meadows, avoidance of the application of certain fertilizers and pesticides on grassland.

Suggestions for better integration of environmental and agricultural policies

Except for a few differences in emphasis and viewpoint, the case studies suggest seven major steps for better integrating environmental and agricultural policies:

i) Regular Monitoring and Evaluation of State of Environment

A regular monitoring system concerning the ecological situation, namely levels of contamination of ground and surface waters, residues of plant nutrients and pesticides in food, erosion and run-off, should be in effect at a national, regional and local scale. It is a necessary precondition for any examination of cause-and-effect relationships and for the implementation of regulatory policies.

ii) Full use of Profitable Low Input Practices

There appear to be many good opportunities to improve production practices which combine a reduced use of chemicals, a higher profitability and a better environment. Examples are fertilization according to plant up-take, integrated pest management and reduced tillage. The current extent of unprofitable overuse of agro-chemicals and, hence the opportunity for cost reduction is substantial on many farms.

The further improvement and diffusion of such profitable low input practices is a challenge both for research and extension. Farm advisory services should put high priority on a better propagation of recent innovations in this field.

iii) Decoupling of Income Support from Price Protection

A gradual elimination of price support and a reduction of market interventions would improve the balance of supply and demand on agricultural markets and make more financial resources available for direct income support. A reduced price protection would reduce incentives for highly intensified and specialised production systems, and, thus, contribute to environmental protection. Moreover, direct income transfers could even be combined with ecological necessities by appropriate definition of the conditions under which farmers are eligible for such transfers.

Decoupling is recommended in both the German and the United States studies. The United States report suggests, in particular, that this would reduce government expenditure and be neutral for consumer welfare and farm incomes. The Swedish study also discusses the option of decreasing product prices to the generally much lower world market level and introducing direct payments. However, the study finally tends to give priority to maintaining general price supports and subsidising alternative land uses, including compensation of income losses resulting from limits on production in pollution sensitive areas.

Generally, it is expected that a reduction of product prices will result in a decline in the use of agro-chemicals because profit maximising optimum input levels will be lower than at the higher price level. Empirical estimates underline that this is a realistic assumption. The elasticity of response, however, will vary according to location and farming system.

The response may be quite elastic on locations with low productivity. Farmers in areas of high soil fertility, on the other hand, including those in some of the study regions, may tend to maintain high input intensities in spite of product price reductions. In any case, a drastic decline of producer prices would cause a drop of gross margins per hectare. For a number of farms, the reduced gross margins might no longer suffice to cover fixed costs, which may result in an abandonment of land or transfer of land to other farms. Whatever the specific form of structural change may be, it is rather likely that it will result in a lower level of intensity of land use.

Yet, parallel to extensification in response to a decline of producer prices, there may be forces towards a continued intensification. One is further intensification of farms which so far had not reached their optimum use level of agro-chemicals. Another is ongoing technical progress tending to shift the production functions upwards towards higher optimal input levels. Historically, such compensatory forces have sometimes offset the extensification effects of reduced product price on agro-chemical price ratios. However, this may no longer be so in case of even greater price reductions. Indeed, it is not unlikely that large price reductions will result in increased fallow, such as the complete abatement of input use in some locations.

iv) Better Use of Existing Production Control Policies for Environmental
   Protection

Recent policy initiatives in the United States, Sweden and Germany tend
to favour quotas on production or land set aside over the elimination of price
support. Various proposals are made in the case studies as to how such
policies can be better implemented in order to realise given environmental
targets. Drawbacks of production quotas, such as long-run inefficiencies due
to rigidities and price increases to consumers are also discussed.

Concerning fallow and land set aside, the studies arrive at a rather
critical assessment. Even in the long run, the ecological value of fallow is
limited, unless it is located in areas of specific ecological value and unless
the consequent maintenance practice is ecologically supervised.

Land set aside is advantageous in regions with a high priority for
afforestation or re-establishment of biotopes. The Swedish study gives
examples for such situations. A premium for the set aside of agricultural
land, however, may not in all cases be a good long-term approach to reduce
agricultural surpluses. The introduction of extensive forms of land use,
possibly in combination with structural change and farm growth might be a
viable alternative which could be motivated by a reduction of price support
and alternative forms of income transfers.

v) Taxes on Fertilizer and/or Pesticides, Combined with
   Location-Specific Regulations

Taxes and levies on certain inputs or agricultural practices may be
necessary in cases where an elimination of producer price support is either
not politically feasible or insufficient to meet given environmental targets.

Yet, due to substantial interregional differences in yield response to
input variations as well as of environmental hazards, taxes may have to be
combined with location-specific regulations. Major problems of nitrate
leaching occur in regions with high livestock density. Therefore, the German
and Swedish studies favour direct location-specific stocking limits and
incentives for storage and interfarm exchange of manure. If such regional
measures are combined with a general tax on nitrogen fertilizer, such a tax
would be based on mineral as well as on organic fertilizer. This is proposed
in the German study. Of course, regulatory measures, taxes and levies should
only be applied insofar as the desired adjustments of production systems are
considered to be part of the farmers' responsibility for the society's
ecological goals. Additional adjustments justify compensation.

Generally, the elasticities of input response with respect to changes
in input prices tend to be similar to those with respect to opposite changes
in product prices. Yet, the resulting income losses are considerably smaller
in the case of taxes. This implies that input taxes tend to be preferred over
a reduction of producer price support if drastic changes in price ratios are
required in order to induce an appreciable reduction in input use.

vi) Stricter Environmental Regulations and Enforcement of Existing
   Legislation

All three countries already have a substantial legislation aiming at
the protection of environment and at the ecological control of agricultural
production systems. Administrative or operational difficulties with respect
to the full implementation of this legislation should be overcome with high
priority.

Insofar as current regulations are insufficient, the legal framework
will have to be further developed which ensures that excessive input levels
are avoided, hazardous pesticides are banned and harmful practices are
eliminated. This will require operational ecological norms and targets.

vii) Polluter-Pays Principle in Agriculture - Financial Incentives for
    Positive Ecological Contributions

According to the Polluter-Pays Principle, the farm population will have
to accept such restrictions which aim at avoiding environmental damaging
without being entitled to financial compensation in all cases. It is
difficult, however, to define tolerance limits for pollutants and to trace the
originators, at least in cases of non-point pollution. Consequently, it will
be an important task for research, administration and political bodies to
specify controllable ecological targets to be met by agricultural practices in
order to avoid environmental hazards. Moreover, politicians will have to
define the borderline between adjustments which farmers have to accept as part
of their social responsibility for the environment ("orderly agriculture" in
German legislation) on the one hand and additional activities which facilitate
long-term ecological benefits and for which the society is ready to compensate
the farmer, on the other.

Chapter 3

INTENSIVE ANIMAL HUSBANDRY AND MANAGEMENT OF ANIMAL MANURE

P. Rainelli (1)

INTRODUCTION

This chapter summarises the policy area studies on intensive animal husbandry and the management of animal manure in France and The Netherlands. Each study consisted of a review of the national situation followed by the detailed examination of the effects of current agricultural policies and feasible policy options on the environment. Information from other countries is also included.

Development of intensive animal husbandry

The sharp growth of intensive livestock production, which occurred in the 1950s for poultry and the 60s for pigs and vealer calves, may be ascribed to several economic and social factors. During this period of economic growth, urbanisation and rising incomes led to major changes in lifestyle, resulting in higher demand for meat products. Poultry and pigmeat, which are less expensive than beef, filled the gap.

On the supply side, it was the more dynamic farmers who, lacking the necessary structure for land-based development, specialised in poultry and pigmeat production. This specialisation and expansion of intensive farming was also considerably abetted by the powerful co-operative and private agro-food complexes which supply animal feed and produce meat products.

It should be noted that the expansion of intensive farming in Europe is only indirectly connected with the Common Agricultural Policy since there is

---

1.   This summary of policy area studies was prepared by P. Rainelli, in consultation with a study group including the authors of the two studies, representatives of the Dutch and French governments and a representative of the OECD Secretariat. The Dutch policy area study was written by Grontmij n.v. and the French study was prepared by P. Rainelli.

little direct Community assistance for pork and even less for poultry and veal production. The Common Agricultural Policy and related tariff arrangements, however, do have a direct effect on the price of feed inputs, the general demand for meat products and, hence, the nature, location and intensity of intensive animal husbandry.

Pollution problems

The very sharp growth in the number of intensive animal husbandry farms in certain areas in France (Brittany) and The Netherlands (in the centre, south and east) has had an appreciable impact on the environment. In some cases, the adverse effects are compounded by the presence of effluents from dairy or beef production units in addition to those from poultry and pig farms. It should be noted, however, that pig manure presents a more serious problem as it is more liquid than manure from poultry production.

There are several different ways in which intensive livestock production units can pollute the environment. First, the mere presence of such units creates local problems through noise and smell, while the visual landscape is also affected since it is spoilt by the buildings and facilities. Second, specific problems arise from the storage and spreading of effluents, either through run-off or leaching of manure components into the soil and also through the emmission of acidic gasses into the air.

Apart from neighbourhood or esthetic problems, manure pollution can affect health, specific local ecosystems and ecosystems in general. Most of the effects on health concern the deterioration of the potable water supply, whether surface or ground water, through nitrates which may be converted into carcinogenic compounds harmful to babies and pregnant women. Consequently, the European Economic Community has laid down a guideline of 25 mg of NO_3 per litre and a permissible level of 50 mg for potable water supplies. It should also be noted that nitrates are harmful when ingested by animals. Nitrites as micro-pollutants are also hazardous as such. Pathogens in manure have also been suspected of contaminating shellfish.

Local effects on ecosystems of intensive animal production include ammonia emissions, which are toxic to neighbouring vegetation, including conifers. Soil contamination by trace elements, chiefly copper and zinc, which occur in animal feed and therefore also in manure, has been observed. Following these observations, regulations to reduce the quantity of zinc and copper in animal feed have been introduced in several countries. The next most important nutrient after nitrogen, phosphorus from manure remains in the soil since plant uptake is low relative to the quantities supplied. In the long run, phosphorous accumulation may lead to crop yield reductions.

Diffuse effects on ecosystems include losses of ammoniacal nitrogen through volatilisation and ensuing acid precipitation, which in turn acidifies the soil in other places. Nitrogen and phosphorus may be transported by run-off water and leakage from improperly constructed manure tanks can lead to the eutrophication of slow-running water. This affects the ecosystems in rivers and also makes the treatment of potable water more complex. Eutrophication of estuarine and coastal waters may also occur, with the possibility of algal blooms.

Table 3.1 sets out the factors associated with intensive animal production which cause pollution, their effects and consequences as pollutants.

Table 3.1

POLLUTION FROM INTENSIVE ANIMAL HUSBANDRY: FACTORS, EFFECTS AND CONSEQUENCES

Factors	Effects	Consequences
1. Siting of the buildings	Noise, appearance, smell	Neighbourhood and aesthetic problems
2. Run-off from Spreading operations Nitrogen infiltration	Smell, Production of nitrates	Neighbourhood and tourism problems, Drinking water quality
3. Leaching of organic matter, nitrogen and phosphorus	Oxygen absorption in flowing water, Eutrophication of slow-running water and possibly estuarine and coastal water	Disruption of river ecosystems, Potable water quality
4. Transport of pathogens	Bacterial contamination of shellfish	Effects on human health and shellfish marketing
5. Atmospheric emission of ammonia	Acidification of the environment through depositing and leaching	Toxic effect on plants, Effects on the soil
6. Manure storage facilities capacity	Spreading at sub-optimal periods, seepage from tanks	All of above

Approaches in other countries

Portugal has reported environmental problems as a result of the presence of copper and zinc in the feed supplements used for livestock breeding. The European Community has set a standard for cadmium in feedstuffs and compound feeds. Scandinavian countries have had to introduce stricter regulations on manure spreading in winter, given the problems of runoff from frozen ground.

To control and prevent pollution problems associated with animal manure, several countries have introduced storage capacity and spreading regulations. In Denmark and in some German and Austrian Laender, regulations require that liquid manure be ploughed or harrowed into the soil within 24 hours unless it is applied to a crop or a pasture. In the United States, a licence is required before any form of waste from a large animal handling facility is disposed of into a water course. Odour problems have led to the passage of legislation in Canada and the United States which prohibits the location of animal husbandry units in certain areas.

In Denmark, a special set of regulations have been introduced with a view to reducing nitrate pollution by 50 per cent and phosphate pollution by 80 per cent before 1992. These include a requirement for nine months storage and for 45 per cent of each farm to be under green crops or pasture in autumn 1988. The autumn green crop requirement will be increased to 55 per cent in 1989 and 65 per cent in 1990. Farmers are also required to prepare manure management plans which indicate that they will not cause excessive pollution from their spreading of both manufactures and animal manure. In countries where there are restrictions on manure spreading in winter, it is usual to require between six and nine months manure storage capacity.

Programmes of grants to assist farmers to upgrade their manure storage facilities have existed in several countries, notably Sweden and Norway. These programmes, however, were only introduced during the transitionary period when new storage regulations were introduced. Finland is currently subsidising the cost of upgrading liquid manure storage facilities. On the other hand, European Community policy now excludes the allocation of grants or concessions to poultry farmers for production increases, and direct subsidies of any kind to farmers who run more than 550 pigs.

To reduce the intensity of production and enhance environmental values a wide variety of production limitations are also being introduced. One example of this is the requirement in Finland that permission be obtained from the National Agricultural Board to establish a holding above 60 beef cattle, 200 eight-week old pigs, 1 000 laying hens or 30 000 poultry. Along similar lines but with the additional aim of protecting rural industries an act was passed in Austria in 1983 to require farmers to obtain permission to keep more than 400 fattening pigs, 50 breeding sows, 130 fattening calves, 10 000 laying hens, 22 000 fattening hens, 22 000 pullets, or 12 000 turkeys. When more than one of these types of animal is kept a proportional adjustment is made to these limits. In Japan, in regions where odour and other environmental problems associated with intensive animal husbandry exist, assistance is given to farmers to encourage them to relocate to areas where their facilities will have less impact.

The general conclusion from the above is that generally countries are beginning to adopt animal manure pollution control and prevention policies in a manner which is within the spirit of the OECD's Polluter-Pays Principle. Under this Principle, the costs of any agricultural pollution prevention and control activities should be paid by farmers. An exception is made, however, to assist farmers during the transitionary period when new regulations are introduced.

French situation (the case of Brittany)

Compared with other Community agricultural sectors, intensive livestock farming is less protected and more vulnerable to world market fluctuations. France is a leading poultry producer and in 1986 it returned to the 1982 production level after a lull in 1984. Owing to improvements in the export market, duck and turkey production is showing sharp growth, albeit to the detriment of chicken production. Generally profits from poultry production have been good.

The situation concerning eggs for consumption is critical, however, since output has failed to recover to the 1982 level and moreover has markedly diminished since the end of 1984. At the same time producer prices are falling appreciably and egg imports now exceed exports. During 1986 egg prices fell by 15 per cent.

Pork production rose slightly in 1986 after falling in 1985. French domestic production, however, is increasingly unable to meet demand and the country was only 78 per cent self-sufficient in 1986,. Whilst pig feed prices fell by 6 per cent, producer prices fell by 9 per cent in 1986, causing the gross margin per pig produced to fall by about FF 50.

About 40 per cent of the intensive farming production in France comes from Brittany. The region supplies half of France's total pig production and one-third of poultry production. Since farm size is small in Brittany the disposal of animal waste is a problem as farmers often do not have sufficient land over which they can spread the manure without causing pollution. This is compounded by the fact that Brittany also happens to be the leading dairy producer region in France.

Pollution problems in Brittany

Owing to its geological features, Brittany does not have many underground water sources and they only account for 20 per cent of the total domestic water supply. The remaining 80 per cent of domestic water supplies are derived from river reservoirs. Consequently the run-off of manure into surface water is the principal environmental problem.

Animal manure is usually spread over farm land for its fertilizer value, although this practice closely depends on the origin of the manure (cattle, poultry or pigs) since the composition differs. Fertilizer value is also governed by the nature of the liquid and solid feedstuffs used, cleaning and storage methods and is often not fully appreciated by farmers. In Brittany for instance, in the areas with a very high pig density, surveys show that barely one third of producers are aware of the fertilizer value of manure, and that they consequently rely heavily or exclusively on inorganic fertilizers. Similar situations arise in Sweden.

Soil nature and condition, types of crop and weather conditions govern the spreading of manure. Detailed surveys show that 40 to 50 per cent of the area is unsuitable for spreading owing to the slope of the land, hydromorphic soil or location next to rivers and streams. Areas available throughout the year for manure spreading cover at most a quarter of the total farming area.

Map 3.1 summarises the average nitrogen load at cantonal level throughout Brittany supplied as well as nitrate concentration in rivers in 1986. In the early 1980s only Northern Finistère showed nitrate levels above 50 mg NO_3 per litre whereas there are now 5 other areas which have exceeded this threshold and 21 others with levels higher than 40 mg. It should be noted that on the northern coast, especially in Finistère, intensive farming is carried out alongside market gardening, which requires large inorganic nitrogen inputs. As a result, denitrification plants have had to be installed for potable water supply.

Map 3.1

Average nitrogen loads from farm effluents (N/ha usable area)

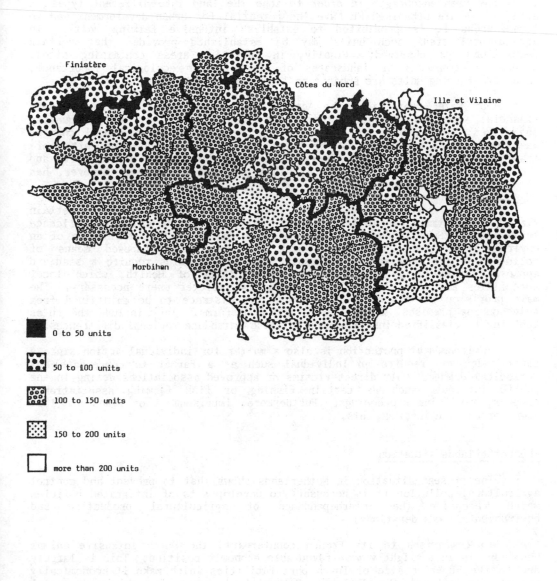

0 to 50 units

50 to 100 units

100 to 150 units

150 to 200 units

more than 200 units

Environmental protection measures in Brittany

Both general and specific regulations are used to control the disposal of animal manure in Brittany. Under land-use regulations, the siting of intensive farming units must not disrupt the environment or the neighbourhood. To this end, the drawing up of land-use plans at communal level has been encouraged in order to zone the land into different types of area. Some are urbanised or have the potential for urban development and in these areas it is prohibited to establish intensive farming units. In agricultural areas such units may be established provided that certain regulations are observed. Finally, in sensitive areas containing sites, natural environments or landscapes of aesthetic or ecological importance, intensive farming units are subject to licensing requirements.

For water protection, the Act of 16th December 1964 setting up basin financial agencies provides that each agency may set water quality objectives. Under this Act any discharge leading to environmental changes must be licensed and taxes may be levied on pig production units with more than 56 pigs. The tax system takes into account the size of the pig farms and the spreading quality of the manure. This proposed tax system, however, has not yet been introduced.

Specific regulations exist for classified installations above a certain size. For instance, above 450 pigs, 250 calves or 20 000 chickens a licence is required. Such licences are granted only following the examination of an environmental impact assessment which considers the proposed means of pollution control and manure disposal. The health rules require a standard approval document to be issued by the Ministry of Health, which local authorities may adapt by making requirements stricter where necessary. The main provision in this approval concerns the distance to be maintained from neighbouring premises, water courses and fish farms. They include the rules applying to classified installations but also introduce regional differences.

Environmental protection is also a matter for individual action since a court judge may require an individual such as a farmer to compensate any ecological damage. Only direct victims or approved associations acting in the public interest, such as certain fishing or fish farming associations, however, may bring proceedings. Furthermore, imprisonment or fines are laid down for certain infringements.

The Netherlands situation

The present situation in Netherlands shows that to prevent and control agricultural pollution it is necessary to develop sets of integrated policies which recognise the interdependance of agricultural production and environmental considerations.

In comparison to its French counterpart, the Dutch intensive animal husbandry is in a slightly more favourable economic position. This is largely due to the superior size of Dutch port facilities which make it economically optimal to offload the European Communities' quota for imported cereal substitutes such as cassava in The Netherlands. Thus, while the nominal rate of protection for the two countries is similar, Dutch producers have a competitive advantage over French producers. As a result Dutch farmers have taken full advantage of economies of scale and automation with the result that a significant proportion of the Communities' intensive animal husbandry units are now concentrated in The Netherlands. This concentration, however, has had the consequence that the country now has serious manure management problems. Higher labour costs and a greater scarcity of land in The Netherlands have also stimulated the construction of more efficient and specialized production systems.

Trends in intensive farming production over the period 1970 to 1983/84 show very sharp increases in eggs and pigmeat, since Dutch farmers succeeded in anticipating the trends on the European markets and finding new external outlets (see Table 3.2).

Table 3.2.

TRENDS IN INTENSIVE FARMING PRODUCTION IN THE NETHERLANDS

| | Annual production | | | |
| | In thousand tonnes | | Index | |
	1970	1983/84	1970	1983/84
Veal	104	165	100	159
Pigmeat	700	1308	100	187
Poultrymeat	369	512	100	139
Eggs	271	645	100	238

The drop in poultrymeat output which occurred after 1982 was made up by 1985 and the following year, a 2.4 per cent growth was recorded. On the other hand, egg production between 1984 and 1986 showed a drop of about 10 per cent, although this has not prevented The Netherlands from continuing to supply 40 to 45 per cent of the world export market. In the case of pigmeat, The Netherlands is the leading European exporter, alongside Denmark. In 1986 both countries showed the highest growth rates in pig meat exports at 4.9 and 4.7 per cent, respectively.

Pollution problems

The rapid growth of intensive animal husbandry in The Netherlands has led to an approximately 30 per cent increase in farm effluent between 1970 and 1984, when total output reached 98.5 million tonnes. This figure has now risen to over 100 million tonnes. Production has also become more intensified. Out of the total 70 million tonnes of manure produced in The Netherlands in 1970 11 million tonnes were produced by intensive farming. Whereas in 1984 20 million tonnes of farm effluent were produced by intensive units. Spreading of this effluent has doubled the phosphorus and nitrogen input and tripled potassium input to the soil. There are limits on the total amount of effluent which may be spread per hectare in The Netherlands and, consequently, as most farms have insufficient land to spread the manure they produce, there is now a considerable manure surplus. Table 3.3 indicates the volume of effluent used locally and the extent of the manure surplus by type of production.

Table 3.3

EFFLUENTS SPREAD LOCALLY AND SURPLUS EFFLUENTS BY
ORIGIN IN 1984 (in million tonnes) IN THE NETHERLANDS

	Cattle	Pigs	Poultry	Fattening calves	Total
Local use	67.1	14.6	2.6	1.7	86.0
Surplus	7.5	7.5	2.0	1.0	18.0

Since cattle are put out to graze for long periods, the corresponding share of surplus effluents for cattle only totals 11 per cent. Conversely, the surplus from pigs is 52 per cent, fattening calves 60 per cent and chickens 80 per cent. Intensive animal units overall account for 30 per cent of the total effluents produced, but almost 60 per cent of the surplus. The problem, however, does not arise to the same extent in all provinces and is most acute in Gelderland, Brabant and Limburg, where the density measured in numbers of mature cattle per hectare under crop totals 9.1, 11.7 and 10.2, respectively, whereas the national average is 4.4.

The Netherlands and Brittany are of similar surface areas. In The Netherlands, the area under crop is just over 2 million ha and in Brittany, 1.84 million ha. Unlike Brittany, in The Netherlands, ground water provides two thirds of the total potable water supply and as an appreciable proportion of the land is sandy, the leaching of nitrate into ground water creates a health problem. At present, however, the limit of 50 mg of nitrate per litre is only rarely exceeded. But as the nitrate content in soil and water is steadily increasing, it is estimated that 25 per cent of potable ground water supplies are likely to reach or exceed the limit within the next few years.

Apart from water quality problems, ammonia released from farm effluents contributes 20 per cent of the total Dutch acid deposition from the air. In areas where intensive animal husbandry is highly concentrated, almost two-thirds of the trees have been moderately affected by acid deposition, 30 per cent severely affected and only 3 per cent remain unaffected. Moreover, surface waters are becoming more acid and, already, 30 per cent of such water sources have become highly acid, reducing biological richness and increasing fish mortality in rivers.

Environmental protection measures

Given that surplus manure in The Netherlands now totals 20 million tonnes and, although local situations may differ, the Dutch government has decided to tackle the manure surplus problems through several different approaches. These include the:

-- reduction of the mineral content of manure by reducing the mineral content to feedstuffs;

154

-- stimulation of the development of manure treatment and marketing arrangements, including export arrangements;

-- increasing the marketing opportunities for manure by means of quality improvement; and

-- introducing regulations to control the production, spreading and removal of manure.

Under legal measures jointly drawn up by the Ministries of Agriculture and the Environment a first attempt was made to restrict the further expansion of production facilities and livestock numbers in areas showing surpluses (as from 3rd November 1984). Despite this ban, pig numbers increased by 30 per cent between 1985 and 1987, which suggests that it has proved to be either politically or administratively impossible to enforce this ban on production increases.

Following the ban, it was decided to reduce progressively the quantity of manure which may be spread on a farm. This is done by restricting the quantity of phosphate which may be spread per hectare. The initial restrictions are equivalent to a general restriction of four animals per hectare. As yet, for a variety of administrative, political and technical reasons, no attempt has been made to introduce restrictions based on the quantity of nitrogen which may be spread per hectare.

As far as spreading is concerned, the permitted limits of P_2O_5, which may be spread per hectare, take into account soil types (sandy land is an important factor) and the type of crop or plant cover as well as the phosphate saturation level (Table 3.4).

Meanwhile a set of measures designed to promote good animal waste management have been introduced. These include a requirement that farmers prepare manure management plans which include an individual accounting system, the figures produced are then entered in a national databank. These manure management plans are used to enforce the limits set out in Table 3.4 and are also expected to encourage farmers to pay greater attention to the fertilizer value of manure and, consequently, purchase less mineral fertilizer. Advisory services are being strengthened to encourage this development.

In its search for solutions to the manure surplus problems, the Dutch government has given full recognition to the Polluter-Pays Principle and accordingly introduced a general levy on animal feedstuffs and also a levy on surplus production. The general levy applies to all livestock feed manufacturers and will be used to meet research costs associated with manure surpluses and advisory services. The special levy on surplus production of manure is only being imposed on farms with a manure surplus. The size of this surplus levy increases with the quantity of manure produced (Table 3.5).

Table 3.4

MAXIMUM PERMITTED APPLICATION OF PHOSPHATE FOR ANIMAL MANURE
IN KG PER HA PER YEAR

Phase	Period	Grassland	Fodder maize land	Arable land *
1	1 May 1987 - 1 Jan 1991	250	350	125
2	1 Jan 1991 - 1 Jan 1995	200	250	125
3	from 1 Jan 1995	approx 175	approx 175	125
4	from approx. year 2000	final standard	final standard	final standard

* Average, standards vary with soil type

Table 3.5

THE SYSTEM OF SURPLUS LEVIES

Production in kg phosphate per ha per year	levy per kg phosphate per year
0 - 125	no levy
126 - 200	Gld 0.025
more than 200	Gld 0.50

There are various possibilities for the surplus levy of Gld 0.50 to be reduced, for example, when dry chicken manure is produced. The funds raised in this way will be used to promote the efficient disposal of manure. Among the facilities to be financed by the surplus levy are a national 'manure bank', central manure storage facilities and manure treatment, and, also, processing or destruction plants. Feed manufacturers and importers are also being encouraged to build such facilities and they are seeing this as being an activity which is necessary in their own interests.

The surplus levy is expected to affect about 43 000 production units at an average cost of Gld 990 per farm. In addition to paying this levy, farmers who have surplus manure will also have to meet the full cost of collection and disposal of manure from their farms.

In addition to all the above measures, several regulations govern the protection of woodland and areas sensitive to ammonia emissions. Other regulations to prevent spreading on frozen and snow covered land come into effect in 1988.

Suggestions for the better integration of Agricultural and Environmental Policies

The experiences reflected in the policy area studies in France and The Netherlands coupled with additional information from other countries suggests seven major points which are relevant to the integration of agricultural and environmental policies.

i) There are several opportunities for the more efficient use of animal manure as an agricultural fertilizer in a manner which both increases farm income and decreases agricultural pollution.

These opportunities include the strengthening of advisory and research activities. Another is the involvement of the suppliers of animal feed in the development of improved disposal methods, processing plants and markets for animal manure. The use of animal manure as a fertilizer is often inefficient and can be enhanced through the strengthening of advisory and research activities. Opportunities also exist to require farmers to prepare manure and/or nitrogen management plans both as a mechanism to help enforce regulations and also as a means to make them more aware of the fertilizer value of their manure.

ii) The relative prices of feed and other inputs influence and can be used to manipulate practices which cause agricultural pollution in an environmentally favourable manner.

The composition of animal manure also depends on price ratios between feed and other inputs. Lower cereal prices provide economic incentives for farmers to modify the composition of manure since more cereals would be used. But lower cereal prices could also encourage growers to begin intensive animal husbandry in a manner which leads to the relocation of intensive husbandry units away from The Netherlands. On the other hand, mineral fertilizer prices, which determine the opportunity cost of manure, also influence manure disposal practices. Higher fertilizer prices would make manure banks more operational by expanding their zone of attraction. It would also make manure conversion operations (extraction of fertilizer nutrients) more attractive and encourage farmers to make better use of manure as a fertilizer.

iii) In developing integrated policies, it is important to give full consideration to regional differences in physical, ecological and social conditions.

Ecological and geological conditions vary among and within countries. Consequently, the standards needed to ensure environmental protection vary from region to region. In large countries, national standards are not always appropriate and it is necessary to have regional standards. In France, for example, the management of animal manure is a regional problem because the intensive farming production is mainly located in one region. In The Netherlands, however, the management of animal manure is a national problem because there is a considerable surplus throughout the whole country in reference to the available land. The present situation in The Netherlands shows that to prevent and control agricultural pollution it is necessary to develop sets of integrated policies which recognise the interdependance of agricultural production and environmental considerations.

iv) The adverse affects of agricultural pollution from the inappropriate disposal and storage of animal manure can be efficiently reduced by applying the Polluter-Pays Principle to agriculture.

To achieve maximum effect policies aimed at a reduction of the effects of the animal waste surpluses must be consistent with the Polluter-Pays Principle. This is because air and water are common-property resources and, as a result, the market misallocates them in a manner which leads to their over exploitation. The Polluter-Pays Principle by internalising pollution prevention and control costs leads to a better allocation of resources. In the short term the costs of pollution control are borne by the producers. But in the medium term if this has an undesirable effect on profits farmers will be forced to transmit these extra costs to consumers. In the long run, the consumers will pay more for their porkmeat, but they will have a better environment.

The Polluter-Pays Principle may be applied in the case of animal manure in assessing levies on feed, on surplus manure, on storage and waste transport. The available funds can be used to finance processing plants and manure banks. But the implementation of these levies raises theoretical and empirical assessment difficulties, mainly because it is difficult to identify the exact extent to which each farmer contributes to the problem. In such situations the pragmatic solution is to define the necessary practices and require farmers to pay their costs. Off farm pollution control activities can then be financed via a levy on all producers.

v) The effectiveness of regulations in reducing agricultural pollution is heavily dependent on the degree to which they are enforced.

Governments have a major role to play by laying down and enforcing strict rules to prevent excess fertilization through the over-application of manure with attendant ammonia release and leaching problems. It is also necessary to prevent spreading on porous soils at unsuitable times, when soils, for example, are frozen and the risk of run-off is high. Such measures should be accompanied by the obligation for each farmer to have adequate manure storage capacity. For new installations the requirement that farmers have to obtain permission to build or expand their facilities is one way of ensuring that adequate facilities are built.

A problem which remains, however, is the enforcement of regulations. Experience reveals that unless standards are enforced, they have very little effect. The enforcement of regulations is also a necessary precondition to the successful application of the Polluter-Pays Principle to agriculture.

vi) Levies and taxes are likely to be more effective if the money collected is used to finance agricultural pollution control and prevention activities.

Generally, farmers are opposed to the introduction of levies and taxes. Experience in The Netherlands and, also in Sweden where taxes on mineral fertilizers and pesticides exist, has revealed that political opposition to these taxes is much less if the money collected is placed in a special fund to help overcome the pollution problem. In The Netherlands, acceptance of the levy is high as the money collected is used to finance research and extension directed towards reducing pollution from intensive animal husbandry.

Levies and taxes, however, have the problem that they are insensitive to input and output price fluctuations and, hence, on their own are likely to often prove ineffective in adequately controlling pollution. Usually additional regulations which set minimum standards for equipment and limits on the amount of manure which may be spread are also necessary.

vii) International trade and tariff arrangements can have an important influence on the location and intensity of agricultural production and through this, the severity of agricultural pollution problems.

International tariff and trade arrangements have had an important influence on the location of intensive animal husbandry. Although the level of financial assistance and protection for this industry is low, it is clear that the tariff arrangements for imported feed stuffs such as cassava and corn glutten have led to the dramatic increase in intensive animal husbandry in The Netherlands. Pollution from intensive animal husbandry would be much less in The Netherlands if either the tariff for feed imports was increased or the subsidies for cereal production were reduced.

Chapter 4

DRYLAND FARMING, SOIL CONSERVATION AND EROSION

Pierre Crosson (1)

INTRODUCTION

This chapter summarises studies for the United States and Portugal
dealing with the integration of environmental policies with soil conservation
policies in areas of the two countries engaged in dryland agriculture (2).
Each study consisted of a review of the national situation followed by the
detailed examination of the effects of current agricultural policies and
feasible policy options on the environment. Information from other countries
is also included.

The United States and Portuguese studies evidence both similarities and
differences in the soil erosion problems the two countries face, in the
policies they have adopted to deal with the problem, and in the relationship
of these policies to agricultural policies generally.

Nature of Erosion Problems

Portugal lacks comprehensive data on soil erosion, but the country's
soils, cropping practices, experimental data, and simple observation indicate
that on a per hectare basis erosion is much lower in Portugal -- probably well
under half as much -- than in the United States. Soils in Portugal, however,
typically are much thinner than in the United States. Consequently, the lower
per hectare erosion rate in Portugal does not necessarily mean that the threat
of erosion to soil productivity is less in Portugal than in the United States.

---

1. This summary of two policy area studies was prepared by Pierre Crosson,
    in consultation with a study group including the authors of the
    studies, representatives of the Portuguese and United States
    governments and a representative of the Secretariat. The Portuguese
    study was written by A. G. Ferreira and the United States study by
    P. Crosson.

2. Dryland agriculture as used here is synonymous with rain-fed
    agriculture.

Despite the difference in per hectare erosion rates the erosion problems in the two countries have an important common characteristic: the off-farm damages of sediment are usually greater than the damages of erosion to soil productivity. The evidence for this in the United States is particularly strong. Studies using the Productivity Index model developed by soil scientists at the University of Minnesota, the Erosion-Productivity Impact Calculator model developed by the Department of Agriculture (USDA) and by Crosson at Resources for the Future concur in showing that if present rates of cropland erosion continue for 100 years, crop yields at the end of the period will be 5-10 per cent less than they would be otherwise. Using crop prices of the early 1980s and a 5 per cent discount rate, Crosson calculated the annualised present value of this production loss to be $500 million - $600 million. A study by the Conservation Foundation estimated the off-farm costs of sediment damage -- shortened reservoir life, costs of dredging, loss of recreational values, etc. -- to be between $2 billion and $13 billion, with a "best estimate" of $6 billion (1980 prices). By these calculations, off-farm damage exceeds on-farm damage by between 3 and 26 times.

The data needed to estimate erosion damages are quite limited in Portugal, although extensive soil type and soil use mapping has been carried out by the National Centre of Agricultural Planning. Although loss of soil productivity in eroded places is a serious problem in some areas sediment damage in places of deposition in general seems to be a major problem. Sedimentation on fields and in drainage ditches so impairs drainage that severe water-logging occurs in wheat fields with a resultant loss of yield.

Soil Conservation Policies

When Federal soil conservation policies were first established in the United States in the 1930s erosion damage to soil productivity was thought to be the main threat, and reduction of erosion was the principal objective of soil conservation policy. While the USDA now is giving more attention to off-farm sediment damage, reflecting the recent evidence showing that these damages are more severe than the loss of soil productivity, policy continues to principally focus on productivity loss rather than the off site effects of erosion.

Until the 1985 Food Security Act, the main instruments of soil conservation policy were an offer to share with farmers the cost of erosion control practices, technical assistance in the design and implementation of these practices, education about the nature and consequences of erosion, and persuasion of farmers, emphasizing the ethical obligation of each generation to protect the interests of future generations in the productivity of the soil. The guiding principle was to induce farmers to voluntarily adopt soil conservation measures. With the exception of grazing land leased from the Government in the United States, however, regulatory approaches have never been seriously considered as instruments of soil conservation policy in the United States or Portugal. Voluntary contracts between farmers and the Government are used in both countries.

Policy is implemented by two USDA agencies, the Soil Conservation Service and the Agricultural Stabilization and Conservation Service in co-operation with the Soil Conservation Districts, local agencies organised under state laws. There is approximately one Soil Conservation District for every county in the nation. While this system has the strength of bringing local knowledge and initiative to bear on soil erosion problems, it also has created political pressure to diffuse federal soil conservation resources widely around the country. Studies by the General Accounting Office (an arm of the Congress) and the Agricultural Stabilization and Conservation Service indicate that a significant amount of these resources has been used to fund soil conservation measures on land with little erosion potential. These measures have often enhanced soil productivity.

The main elements of soil conservation policy in Portugal also were put in place in the 1930s. Unlike the United States where cropland erosion was seen as the principal problem, however, Portugal considered deforestation and overgrazing to be the main causes of erosion. Accordingly, reforestation and establishment of permanent pasture were adopted as, and continue to be, the principal instruments of soil conservation policy.

In Portugal, there are about 500 000 hectares of communal land owned by parishes or municipalities, almost all of it in the mountainous region north of the Tejo River. The land is used as a source of firewood, for animal grazing, and other miscellaneous purposes. The combined effect of these different uses has been to degrade the vegetative cover and stimulate erosion. Reforestation and pasture establishment programmes on this land are managed by the Forest Service, which now administers 600 000 hectares of communal and state owned land, and also undertakes some reforestation programmes on private land.

As in the United States, soil conservation policy in Portugal relies on measures to induce farmers to voluntarily adopt conservation measures. From 1963 to 1980 the government provided low interest, long-term loans to private landowners who participated in the reforestation programme or who established permanent pastures on their land. The Forest Service provided technical assistance to these landowners and also set regulations to control wood and game harvesting on the land. When the program ended in 1980 some 170 000 hectares of private land had been reforested. By 1986 an additional 55 000 hectares of private land had been reforested under a programme sponsored jointly by the Portuguese Forest Development and Credit Agency and the World Bank.

Yet another Portuguese programme deals with gully erosion in the watershed of the Lis River and the southern part of the watershed of the Mondego River. Gully erosion in these areas had been a major problem, but now is under control, thanks to work which began in 1938 as part of the Communal Lands Reforestation Programme and a water erosion control programme. With the permission of land owners, the Forest Service undertook to plant trees to stabilize the gullies. Landowners agreed not to harvest the trees without permission of the Forest Service and under procedures that prevented re-emergence of the problem.

In other OECD countries, soil conservation policies are also evolving. In Canada and Australia, regulations are being introduced to maintain and protect the long-term productivity of land, notably rangeland areas, subjected by economic pressures to overuse and degradation. Soil conservation research is actively promoted in many countries.

Subsidies to farmers for erosion prevention are a common feature in many countries, examples being grants for the construction of soil contour banks in Australia, the supply of inputs such as trees at subsidised prices in Turkey and cost-sharing programmes in Canada to encourage the adoption conservation practices which reduce erosion and flood damage. Tax allowances exist in Australia for expenditures on soil conservation.

Integration of Policies

Policy integration as discussed here has both substantive and institutional dimensions. The substantive integration issue can be stated as follows: to what extent do macro-agricultural policies take into account their effects on soil erosion, and to what extent do soil conservation policies take account of their effects on agricultural output, income and prices? The institutional issue is the extent to which the exchanges of information and people among agencies actually result in closer harmony among agricultural and environmental policies than would occur in the absence of the exchanges.

Substantive integration

Soil conservation policy in Portugal is poorly integrated with policies designed to support agricultural prices and farm income. This is true also of the United States, although the conservation provisions of the 1985 Food Security Act mark a significant, if still small, step toward closer integration. Since the 1930s both countries have had policies to support crop prices, primarily of cereals, as a way of maintaining farm income.

In Portugal an additional policy objective was to reduce the country's dependence on foreign sources of food. The policy has worked in the sense that farm income has been higher and cereal imports lower than they otherwise would have been. The policy, however, also caused more land to be in cereal production than otherwise would have been the case. Correspondingly, as 23 per cent of Portugal's cultivated area is deemed unsuitable for agriculture, erosion and attendant environmental damages were higher than they otherwise would have been.

In the United States, the erosion consequences of the failure to integrate macro-agricultural policies with soil conservation policies are not clear. Like Portugal, several product price support schemes have and continue to operate in the United States and as a result more highly erodible land and more marginal land is under crop production than would otherwise be the case. Although between 1934 and 1985 farmers under various schemes were given inducements to take land out of production as a way of reducing crop supply, which would tend to reduce erosion, they were not required to retire their most erosion-prone land. The contracts for land retirement were also only for one year, so whatever erosion reduction benefits were achieved in one year could be lost in the next. Moreover, since the scale of commodity programme benefits farmers received was based in part on the number of their "eligible" crop acres, they had an incentive to begin cropping highly erodible land to increase their eligibility to participate in the scheme. In this respect, failure to integrate policies may have resulted in more erosion than otherwise would have occurred.

The Conservation Reserve, swampbuster, sod buster and cross compliance provisions of the 1985 Food Security Act offer promise of closer substantive policy integration than in the past. The objectives of the conservation reserve are to reduce erosion; preserve land productivity; reduce sediment-related water pollution; limit the supply of surplus commodities; protect wildlife habitat; and maintain farm income. The idea that land retirement will contribute to environmental improvement is widely shared in the OECD countries. Under the Conservation Reserve Programme farmers submit bids to the Secretary of Agriculture to take some, or all, of their highly erodible land out of crop production and put it in some low erosion use for 10 years. By April 1987 over 19 million acres had been accepted for conservation reserve, which is more than the USDA had anticipated.

On balance, the Conservation Reserve clearly represents an important step toward better integration of macro-agricultural policies and environmental policies. There can be little doubt that the Conservation Reserve will reduce erosion. Whether it will reduce off-farm sediment damage proportionately is problematical. There are strong spatial and temporal discontinuities between the occurrence of erosion on the land and the occurrence of sediment damage downstream. Consequently, the relationship between erosion control and sediment control is highly uncertain. It is to be hoped that as the Conservation Reserve idea matures and experience is gained with managing it, ways can be found to make the conservation reserve a sharper instrument for achieving the objectives of reduced sediment damage and improved water quality.

Like the Conservation Reserve, cross-compliance is a significant step toward better integration of macro-agricultural policies and environmental policies. Cross-compliance makes eligibility for certain agricultural support programmes contingent on the use of approved conservation practices, notably on highly erodible lands. Unlike the conservation reserve, cross-compliance does not necessarily have to rely on land set aside to achieve environmental objectives and, importantly, it seeks to rectify the environmentally negative impacts of price support. There is no reason in principle why cross-compliance could not be used to induce environmentally desirable changes in pesticide, fertilizer, and irrigation practices, as well as in land management. Cross-compliance thus seems a more flexible instrument than the conservation reserves for achieving policy integration.

The swamp buster and sod buster provisions of the 1985 Food Security Act also offer promise of substantive policy integration because they overcome the tendency of price support to bring more highly erodible and environmentally significant land into production. Both provisions deny farmers access to agricultural subsidies of various kinds if they bring land which has not been cropped during the last five years into production.

From the mid-1960s to the mid-1970s Portugal had a special programme with cross-compliance like features. Its principal aim was to in induce farmers to shift cereal production to land best adapted to that purpose. To this end farmers were offered a mixture of grants and long term, low interest loans, but only on condition that the funds be used for cereal production on suitable land. Subsequently, subsidies were offered for investments which took into account soil potential.

In the United States there is a growing tendency for agencies to define best management practices and then require farmers who wish to qualify for grants for soil conservation, etc., to adopt them. Best management practices have been defined in many areas for conservation tillage, the conversion of cropland to permanent vegetative cover to reduce erosion and integrated pest management to reduce environmental damages from pesticides.

In 1987, a new Portuguese reforestation programme with funding from the European Economic Community was introduced in Portugal. Under this programme subsidies will be provided for forest planting and creation of pastures on both private and communal lands, under the joint responsibility of the Forest Service and the Forest Development and Credit Agency.

Institutional integration

The Portuguese study contains little information about institutional integration. In the United States a number of agencies are involved in agricultural and environmental policies. At the national level the Congress is a principal actor, as is the executive branch, operating primarily through the Department of Agriculture (USDA) and the Environmental Protection Agency (EPA). The Council on Environmental Quality has policy coordinating responsibilities, but little authority. Other agencies, such as the Departments of Interior, Defense and Energy also play peripheral roles. The Department of Interior through its Bureau of Land Management and the USDA through its Forest Service play a significant role in the management of grazed Federal Lands. The dominant federal agencies in the implementation of agricultural and environmental policies are the USDA and EPA. In the formulation of policies these agencies share responsibilities with the Congress.

State and local agencies also are active in the policy process, in many but not all instances operating within advisory or regulatory guidelines set by the USDA and the EPA. Some states have imposed more stringent environmental policies than EPA and USDA. For example, California places more stringent restrictions on use of some pesticides, and Iowa on permissible rates of soil erosion to reduce off-farm sediment damage.

Because policy for use of non-federal lands is a state responsibility in the United States, state and local agencies are particularly active in the implementation of soil conservation policies. Local soil conservation districts, established under state law, are the principal agencies responsible for implementing soil conservation policies. All states have agencies devoted to protection of water quality and other environmental values. To a greater or lesser extent these agencies are involved in environmental aspects of soil conservation policy.

In 1974 the Secretary of the USDA and the Administrator of the EPA signed an agreement providing for coordination of activities of the two agencies. Under the agreement the agencies have undertaken numerous joint research, training and project implementation activities. In addition, the agreement provided a useful mechanism for frequent meetings between top level officials of both agencies where issues of mutual concern could be discussed.

The legislation establishing the Rural Clean Water Programme states that at the national level the Secretary of Agriculture will administer the programme in consultation with the Administrator of the EPA. It is stipulated that the EPA must concur with USDA in selection of the Best Management Practices which under the programme farmers are encouraged to adopt. Day-to-day administration of the Rural Clean Water Programme was delegated to the Agricultural Stabilization and Conservation Service and coordination of technical assistance to the Soil Conservation Service.

Although the potential range of collaboration between the USDA and the EPA is wide, it is fair to say that most of the interaction between the two agencies has been with respect to pesticide policy. In matters touching on agriculture, this was, and is, the EPA's principal concern. It, perforce, was, and is a main concern of the USDA as well. In contrast, the two agencies have had much less to say to one another about soil conservation policy.

The contrast emerges clearly from the United States Country Information Paper. Part II on Policy Areas has a section titled "Intensive Crop Production/Agricultural Chemicals Interface" and another titled "Dryland Agriculture/Soil Conservation and Erosion Interface". The Agricultural chemicals section deals mostly with pesticides. It briefly discusses USDA's role in doing research and extension aimed at providing farmers with economical, less environmentally threatening pest management technology, e.g., integrated pest management. But most of the discussion is about EPA policies in pesticide regulation. Integration of USDA and EPA activities is not stressed, but one can see that it occurs.

The section on soil conservation, however, is all about USDA activities. The EPA is not mentioned.

The main policy area in which the two agencies have common responsibilities is in connection with non-point pollution. As indicated above, section 208 of the Clean Water Act Amendments of 1972 assigns the EPA principal policy responsibility for this kind of pollution. The EPA has chosen to delegate this responsibility to the states, which for the most part have seen the non-point problem as a problem in erosion control. Accordingly, they have relied heavily on the SCS, and state soil conservation agencies, to deal with the problem. The EPA has played a role in this, but a minor one. According to the United States Country Information Paper, EPA has accepted SCS personnel in regional EPA offices to help co-ordinate EPA and USDA work on non-point pollution control policy. In this connection EPA has given USDA people access to its data base on water quality. For the most part, however, interaction between the two agencies on soil conservation policy has been limited.

In the United States a Non-Point Task Force consisting of representatives from the EPA, USDA and the Department of the Interior, as well as other involved Federal Agencies has been formed to facilitate integration. State governments are also represented. In order to implement Task Force objectives, the Secretary of Agriculture adopted a Non-Point Source Policy in November of 1986. This policy directs USDA agencies to place emphasis on protecting water quality from non-point sources of pollution in applicable existing and new programmes. The Soil Conservation Service, for example, is refocusing many of its programmes to place stress on water quality protection and has developed and is now implementing a high priority water quality strategy.

Case Studies

Two areas in each country were selected for more intensive analysis of soil conservation issues, policies, and problems of achieving closer integration of agricultural and soil conservation policies. In the United States the areas were in west Tennessee and east South Dakota. In Portugal both areas were in Alentejo province, one near Beja in the southern part of the Province and the other near Evora in the north.

The case studies generally confirmed the main conclusions reached from a national perspective:

a) In the areas studied, off-site damages of sediment generally appeared to be more severe than on-site damages to soil productivity. Quantitative estimates of the two kinds of damage, however, were available only for the area in South Dakota, where USDA analysts have estimated the value of recreational losses from lake pollution.

b) The policies followed in the four regions to achieve soil (and water) conservation objectives relied strictly on voluntary measures. Farmers were offered financial inducements to adopt conservation practices. Regulatory approaches played no role.

c) Substantive policy integration was weak-to-non-existent in all four regions, although by spring 1987 some land in west Tennessee and east South Dakota was in the Conservation Reserve. In both areas, however, farmers' incentives to put land in the Reserve were weakened because price support for corn made it profitable to keep land in that crop. Eventually a "bonus" was paid to farmers to put corn land in the Reserve. Nonetheless, the situation pointed up the difficulty of integrating policies to support farm income with policies to protect the environment.

d) Institutional integration was not discussed in the Portuguese case studies. The two United States case studies reflected the national situation: little integration of agencies of the USDA and EPA, but considerable involvement of the USDA with state and local agencies with soil and water conservation responsibilities and of EPA with state water quality agencies.

e) The case studies brought out the importance of information about the costs of erosion damages and of the programmes to reduce the damages. Lack of such information seriously hampers effective policy making. Not only is it difficult to judge how much should be spent on damage control and which measures of control are most economical; achieving consensus among those with conflicting interests in land and water management also is made more difficult. Among the four areas studied estimates of erosion damage were available only for east South Dakota. There USDA economists had estimated the value of recreation days lost because of sediment and chemical damage to one of the lakes in the region. Comparable estimates for the other three areas were not done.

Suggestions for better integration of environmental and agricultural policies

For both countries several conclusions emerge relevant to better integration of agricultural and soil conservation policies.

a) The formulation of agricultural policies should explicitly take account of the consequences for farmers' land management decisions, expecially with respect to whether the policies are likely to encourage the farming of more land, or more erosive land, than in the absence of the policies.

b) Soil conservation policy should shift to give more attention to off-farm damages of sediment.

c) Soil conservation policy, both to reduce off-farm damage and on-farm damage, should be more precisely targeted on those places where damage occurs. With respect to control of off-farm damage this suggests more attention should be given to preventing sediment from entering water bodies.

d) Inclusion of "cross-compliance" and "sodbuster" like features in agricultural policies should be encouraged.

e) The environmental protection agencies of both countries should give higher priority to sediment as a major pollutant of surface waters.

f) Collaboration between environmental agencies and agricultural agencies should go beyond agreements to exchange information and personnel. It is important also that the two sorts of agencies jointly and explicitly agree that sediment is a major source of surface water pollution and that they actively collaborate to assure that this fact is reflected in the formulation and implementation of agricultural policies.

g) Although for practical reasons both countries should continue to rely primarily on voluntary compliance of farmers in adoption of soil conservation practices, regulatory approaches shoud not be ruled out in circumstances where off-farm damage is severe and those responsible for it readily identified. In the United States, state agencies probably are best suited to adopt regulatory approaches.

Chapter 5

CHANGING LANDSCAPES AND LAND USE PATTERNS
AND THE QUALITY OF THE RURAL ENVIRONMENT

T. O'Riordan (1)

SETTING THE CONTEXT

This chapter summarises the analysis of three case studies, two in
England and one in Austria. The objective of these studies was to assess the
effectiveness of regulatory and compensatory policies aimed at retaining
traditional patterns of agriculture and woodland management in areas where
such practices might otherwise cease to exist. For both countries
socio-economic changes in agriculture in marginal areas result in loss (due to
intensification) or neglect (because of abandonment) of historical landscape
features. Such an outcome would not just be detrimental from an ecological
and aesthetic point of view. All three case study areas depend to varying
degrees upon tourism for a proportion of their income, so any serious
diminution of landscape beauty would result in loss of revenue and, possibly,
further economic hardship.

The agricultural context for the case studies is common to most western
industrialised countries. Three decades of publicly subsidised agriculture
have impressively modernised the industry. This has created expensive
surpluses of output together with widespread mechanisation and the
intensification of production. The technical capacity for further output
gains is very great, so the problem of overproduction is likely to continue.
The consequence is the impoverishment of less productive farms in the poorer
areas, mostly harsh upland environments or inaccessible wetlands. Yet many of
these farms have long been the mainstay of local economies. Their special
husbandry practices have sustained important ecosystems and retained long
established landscape features.

---

1. This summary of two policy area studies was prepared by T. O'Riordan,
   in consultation with a study group including the authors of the
   studies, representatives of the Austrian and British governments, and a
   representative of the Secretariat. The Austrian study was written by
   W. Puwein and the United Kingdom study by J. Bowers and T. O'Riordan.

Severe economic adversity now faces many of these farms. Their viability cannot be assured by market forces alone since costs have risen far more significantly than market prices for their products. Subsidy in the form of direct income payments or special financial assistance for traditional agricultural practices is necessary if these farm enterprises are to continue. The reasons for the survival of such farms therefore are primarily social and environmental rather than purely agricultural, since for social and environmental reasons, certain patterns of agricultural practice are still wanted from these enterprises. If agricultural overproduction continues to increase, the number of farmholdings dependent largely upon a form of public contract to meet non-agricultural objectives will continue to grow. Member countries will have to decide how much and for how many farms they should provide financial assistance and other special support measures. The policies that underlie these decisions form the central theme of this chapter.

All the case studies examined indicate that post-war agricultural changes have dramatically reduced species diversity, habitat integrity and landscape beauty. Figure 5.1 illustrates habitat losses in the UK. Both the West German and Swedish case studies reveal how intensification and the widespread use of agricultural chemicals have impoverished species numbers and variety and altered wildlife sites. Public opinion in all the countries studied favours the retention of ecological diversity and scenic interest, even if this means increased public expenditure. Redirection of agricultural support payments would be a preferable option.

The importance of land-use planning and the environmental implications of land use, especially in the agricultural context, has been recognised by virtually all OECD Member countries, with the result that land use planning, regulations and incentives aimed at controlling development are widespread and varied. Statutory land use controls are effective policy instruments in France in certain circumstances for ensuring that agricultural development is compatible with environmental policies. Greece and Denmark, for example, have physical planning requirements which, in the latter case, are devolved to twelve regional councils. In Japan, prefects are empowered to introduce land use restrictions in agricultural promotion areas to preserve and improve the environmental amenity of a region. In South Australia, the clearing of vegetation without permission is prohibited. When permission is refused, farmers are compensated for their lost production opportunity and an easement preventing further clearing is entered on the land title.

In the European Community compensatory allowances can be given to farmers operating in small areas affected by specific handicaps and in which farming must be continued in order to ensure the conservation of the environment and the countryside. Community aids are paid and national aids are authorised to farmers who contribute towards the introduction of continued use of agricultural production practices compatible with the requirements for conserving natural habitats and ensuring an adequate income for farmers.

In Switzerland, interest free loans for agricultural improvements and land consolidation now take much more account of environmental considerations. There is a clear policy of subsidising certain desirable agricultural activities. For example, there are payments for the satisfactory cultivation of slopes steeper than 18 per cent to encourage the maintenance of traditional pastures and summer pastures.

Land acquisition and resale to individuals or groups who agree to special environmental management regimes, already a feature of land management in the United Kingdom, and are becoming more common in Canada and the United States.

Figure 5.1

Loss of wildlife habitat in England and Wales since 1945

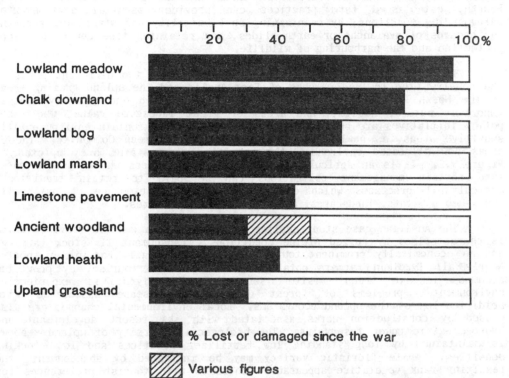

The case studies summarised below concentrate primarily on a particular class of marginal farms. Public support for these farm households rests on four objectives. First, there is a wish to retain viable farm communities in rural areas of high unemployment and depopulation. The existence of farm families helps to retain the fabric of rural society. Networks of farm families provide an economic function by retaining local cultural traditions and servicing tourism. Second, certain patterns of farm management maintain traditional landscapes and archeological features that are increasingly valued by urbanised populations. Third, particular agricultural practices protect, enhance and create ecosystems that are prized for their genetic uniqueness, their special scientific rarity or their unusual biological assemblages. Fourth, established farm practices can provide essential environmental safeguarding functions, by maintaining soil fertility and stability, reducing the hazards of avalanche or earth slides, and retaining tree cover for slope protection and the harbouring of wildlife.

The case studies from the United Kingdom cover two different parts of the country, the lowland marshes of East Anglia and the upland grazing areas of the North Pennines. These locations were selected because they contain important natural habitats and areas of great landscape value, where new policy initiatives are being tried out. Both regions contain environmentally sensitive areas, as newly designated within the European Community. The UK areas now allocated environmentally sensitive area status are depicted in Figure 5.2. It is an agricultural concept where farmers who choose to enter into certain management obligations can be paid to retain traditional agricultural practices which contribute to the maintenance of nationally cherished habitats, landscapes, and archeological features.

The Austrian case study applies to the Tyrol, a western Province which is characterised by rugged Alpine conditions, transhumant livestock farming, and an economically prominent tourist industry especially for winter sports. Nearly all Tyrolean farmers obtain some income from tourism, by providing accommodation or other visitor services. The Tyrol is experiencing environmental problems of forest damage and disease associated with acidification of the upland ecosystems. Local environmental changes are also caused by construction works associated with ski resort development and hydro-electric power reservoirs. The biological diversity of upland meadows is maintained by long standing low fertilizer practices and low stocking densities. While floristic variety may be increased by abandonment, the resulting rank vegetative appearance contrasts with tourist preferences for mown meadows. Alpine grasslands are protected by traditional agricultural practices. But not enough of the forest is protected, ski roads and forestry developments are causing in some cases increased erosion, and forest damage by game and environmental acidification is weakening the resistance of these ecosystems and their capacity to perform their environmental safeguarding functions.

In both England and Austria there is a widely shared view that farmers should be encouraged to meet their land conservation responsibilities through information, advice, financial inducement and peer group pressure rather than by compulsion. In both countries planning controls over the siting of buildings and other developments on agricultural lands do exist, so uncontrolled construction is not possible. Similar controls over changes in agricultural land use apply, however, only to a small number of selected areas.

Figure 5.2

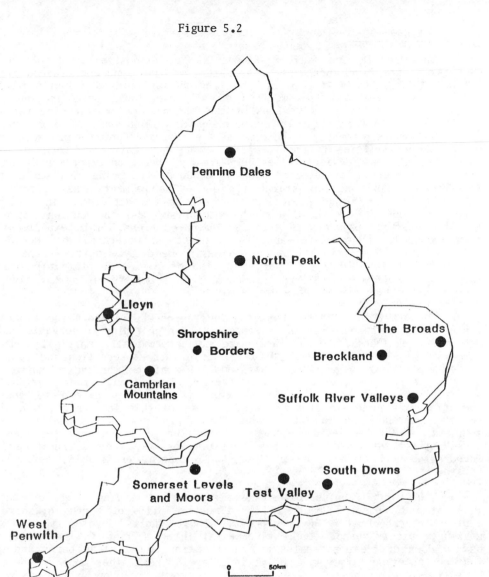

Areas designated as Environmentally Sensitive Areas in England and Wales May 1987.

This designation under European Community legislation permits agricultural departments to pay farmers to retain traditional agricultural practices in order to maintain biologically important habitats, traditional landscapes and to protect historic features.

The case study areas

The Norfolk and Suffolk Broads is an extensive tract of wetland comprising some 20 000 hectares of reed fen, fen woodland and drained grazing marshes in eastern East Anglia. The region is renowned for its wildlife value, its recreational potential and its traditional open grazing marsh landscapes. The livestock regimes of the marshes are experiencing recession due to the rising costs of maintaining adequate drainage and falling livestock prices. This is primarily an outcome of the changing fortunes of agricultural price support policies in the European Community. Only some form of income supplement can ensure viable livestock farming. Under an experimental scheme, livestock farmers in a selected area of 6 000 ha are being paid to continue livestock farming and not start cropping. The payments they receive are sufficient to discourage ploughing and reflect the difference in the gross margin between arable and livestock enterprises and the current level of livestock support payments in the Less Favoured Areas. This experiment has been judged to be successful and, as a result the Government has decided to continue with these payments under the environmentally sensitive area scheme introduced in March 1987. Farmers may also be paid a supplemental grant if they invest in special conservation measures that improve biological diversity and reconstruct cherished landscape features.

The other study in the East Anglian region applies to large estates or groups of farmers where special arrangements for providing conservation advice are being evaluated. These farms are not economically marginal, but the farmers are anxious to establish a mix of nature conservation and landscape interest in their properties. This will be achieved through a "whole farm plan" geared to combining conservation investment and agricultural profitability. The conservation element of the plans extend to tourism, education and business diversification. Some of this investment is designed to return revenue from the sale of small wood products, educational visits and tourism. Preliminary evidence suggests that the retention of a permanent "in house" specialist, paid for by the benefiting estate, plays a crucial role in encouraging and linking conservation to a modern, profitable, agricultural estate.

The North Pennine Dales comprise an outstanding area of habitat interest and landscape beauty in the limestone hills of North Yorkshire and South Northumberland. The region is almost wholly devoted to livestock grazing, notably sheep. The environmental interest lies in the flower-rich hay meadows of the upper dales and in the traditional stone barns and walls that characterise this very special landscape. The heather moor on the upper elevations is overgrazed, perhaps as an outcome of European Community hill livestock support regimes. The present extent of environmentally sensitive area covers four of the Dales in the Yorkshire Dales National Park, but not the moorlands.

The key meadow habitats of the Pennine Dales are granted conservation status as sites of special scientific interest. Such sites are protected by particular legislation that largely ensures that farmers cannot alter their ecological or geological importance, but must be compensated to maintain the scientific interest of such sites where it can be proven that they would lose potential agricultural income by doing so.

The ecological distinctiveness of the Pennine hay meadows is dependent upon low nutrient status, well-timed grazing and hay removal. No artificial fertilizers are added to the floristically diverse grasslands. These meadows form part of an intricate livestock management pattern that has a long history in the Dales. This pattern is changing as sheep fattening practices alter, silage making becomes more feasible, and farm incomes fall. Traditional Dales sheep farming both in the past and in the future has and will depend almost exclusively on support from public funds. Dales farmers are beginning to appreciate this fact.

In the Tyrol, farming is also <u>State</u> supported for public purposes. There are three kinds of support.

i) The Austrian State provides an income support measure to assist small and poor farmers in harsh upland environments.

ii) The Provinces provide compensation for special husbandry practices that retain established alpine meadow management practices and, consequently, a landscape which is attractive to tourists.

iii) The communes disburse special management payments in particular areas to retain traditional practices. These payments mostly involve hay mowing allowances so that meadowlands are cut to maintain the visual appearance of cut grass fields, important for the tourist industry. These payments are financed by the local tourist boards.

Like the Dales and the Broads the most important conservation estate in the Tyrol is safely guarded by special designation. Some 17 per cent of the Tyrolean total area is so protected. This includes nature reserves, rest areas, protected landscapes, moors, lakes, water courses, single trees and tree avenues. The dairy and cattle breeding sectors are supported by guaranteed prices and subsidies. Mountain farmers also receive subsidies for cattle feed, notably domestic grain. The region benefits from various investment grants and special loans that help to modernise agriculture.

Under a Special Programme for Mountain Areas the Austrian Government provides payments to preserve areas that are in danger of being abandoned. Most of these funds are allocated to direct income support, the rest for special assistance in particular regions and for land levelling, forest management and the construction of utility lines (Table 5.1). Yet despite these payments, the net incomes received from all sources by Tyrol farmers still remain below half of the average industrial wage.

Direct payments to mountain farmers can be viewed as a lump sum income supplement for services rendered in maintaining the recreational value of the Tyrolean landscape. The payments reflect the degree of adversity, mostly due to inaccessibility, climate and topography, and the special measures that have to be undertaken. They only apply to small, full-time, farmers; big land owners on above average agricultural incomes are not eligible.

175

Table 5.1

SUBSIDIES AND COMPENSATION PAID TO TYROLEAN FARMERS

	Federal	Provincial
	million Sch.	
Road building	58.2	25.8
Road buildings, special programmes	-	55.0
Cable ways	-	0.9
Electric power	0.1	0.2
Telephone lines	0.9	0.5
Consolidation of land holdings	13.4	8.8
Settlement	0.8	0.5
Subsidies for emigration	-	10.5
Agricultural adjustment	4.6	24.7
Improvements of Alpine pastures	5.7	7.0
Levelling of farm land	0.9	0.8
Regional programme "change over"	11.4	2.3
East-Tyrol, special programme	11.6	8.1
Agricultural river regulations	1.9	7.6
	109.5	149.7

Owners of larch meadows are paid to maintain these woods in an aesthetically healthy condition. Additional subsidies from the local communes are available for special management measures, including the retention of good soil and grass conditions on ski slopes.

In the Pennine Dales case, farmers will continue to need financial assistance if the biological diversity of the hay meadows is to be maintained. This will require special management measures over and above established agricultural practices. These proposals mean that public money is paid to farmers to meet socially agreed objectives. Similarly, in the Tyrol, landscape museumisation may require even higher levels of public subsidy. Some of this new money should assist farmers further to enter the service sector for alpine tourism. In short, these should be restructuring payments.

Policy issues

The three policy area studies examined are illustrative of wider policy considerations. Of prime importance is the essential role of particular patterns of agriculture to maintain publicly valued environmental assets. The state, therefore, has a special role to play in maintaining the agriculture-environment linkage.

Sites or areas of vital conservation importance

Major conservation sites, regarded as having international significance for their landscape typicality, ecological rarity and genetic uniqueness, will always require special protection and particular management. These objectives cannot always be achieved by relying on co-farming measures. Ownership by conservation agencies, wardening and special lease arrangements may be required coupled with strict legal measures to guarantee their total protection.

Established agricultural landscapes

For the more established agricultural landscapes, a mix of management agreements and income support measures should suffice to maintain the agriculture-environment linkage. Income support linked to a requirement for landscape maintenance is most suitable in zones of special hardship where alternative economic opportunities are limited. Management agreements involving special payments are desirable where farm intensification by the next farming generation is likely.

Factors influencing agricultural support payments

The amount of supplementary financing needed will depend on the relative support for "intensive" agriculture available through public subsidy, and the opportunities for non-farm income generation. It is not desirable that future agricultural subsidies should encourage further intensification to produce surplus output. Redirection of such subsidies is preferable so that public money can be channelled into a future agriculture that meets agreed public purposes. Even so, it may still be necessary to provide additional public spending for research and development into more environmentally friendly practices, and for further assistance in farm restructuring so that non-farm income generating opportunities are increased.

Area-based approaches

Area-based or regional schemes for supporting particular agricultural practices are more likely to be cost-effective than farm-specific measures where payments vary from farm to farm. Ideally, an area based scheme should devise arrangements to take into account individual circumstances. Area-based schemes have the attraction of treating farmers reasonably equally, they are more likely to be publicly acceptable, and they are less costly and time consuming to administer. Peer group pressure and neighbourly encouragement can also have a profound educative effect in encouraging farmer compliance for voluntary measures.

Advice and farm heterogeneity

Area-based schemes, however, carry the danger that all participating farms may be treated equally and be given similar advice. This is undesirable as ecological, landscape, historical and management traditions vary to an important extent from farm to farm. Advice is critical, but that advice should be sensitive to farm conditions and farmer circumstances. One way forward is for groups of farmers to have access to a single farm enterprise adviser who would open up avenues to specialised advisory services (in agriculture, wildlife management, new habitat creation, recreation, off farm business generation) and back up financial support. Much can be achieved by encouraging clusters of neighbouring farms to share their ecological and landscape "capital" and benefit from joint marketing ventures.

Indicative land-use planning

The appropriate use of agricultural land needs guidance from authorities for the general regulation of land use. That guidance can best come from indicative land use plans which suggest what areas are appropriate for certain kinds of development and which locations are more suited for conservation investment. The basis of such plans should be set on topography, soils, drainage, geology and ecology together with a sense of history, local culture and amenity considerations. Such plans should provide guidance, not prohibition, and should be drawn up by local authorities on the basis of widespread consultation with interested parties. It would not normally be appropriate for detailed guidance on cropping and stocking to be provided as part of indicative land use planning.

Controls and compulsion

Financial incentives, advice, and indicative land use guides should be given preference over compulsion and control as a first step towards achieving environmental objectives. But for certain highly prized areas, it is desirable that a planning power of last resort should be made available, to ensure that wilful damage is controlled. Where a site is of very special environmental importance, ownership by an accredited conservation agency may be necessary, with scope for leaseback to a willing farmer. Similarly, the possibility of using payments to acquire conservation easements, etc., over land should be considered.

Increasing biological diversity

Biological diversity on modern farmland can be increased if farmers use artificial fertilisers and pesticides more sparingly and more precisely. This can be achieved with new technological advances backed up by sound advice and appropriate management guides. Leaving the edges of fields unfertilized and unsprayed, abandoning small pockets of land and planting wildflowers in unmanaged areas can dramatically improve wildlife diversity and density. Significant opportunities exist at modest cost and little inconvenience to the farmer for the introduction of such schemes.

Suggestions for the better integration of agricultural and environmental policies

Recognising the regional and national differences between the studies, four suggestions for the integration of agricultural and environmental policies emerge.

i) Generally policies which increase the diversity of agriculture and complement agricultural practices which maintain landscape quality are likely to be beneficial to the environment.

The success of integrating agriculture and environmental conservation lies in creating a tradition of earth-sustaining farmers who have sufficient income to remain on the land and become involved in local community life. Such an ethos requires public investment and has a wide measure of public support. Where additional non-farm income can be attained, it is desirable that this should be done by means that are complementary to agricultural practices which maintain landscape quality. In time, a greater variety of national agricultural enterprises can be created. Some farms will intensify, others will shift to part time operations and still others will choose to adopt conserving husbandry. This will open up new oportunities for pursuing an even greater variety of rural policy options, as many farms will no longer be required to produce essential food output.

ii) The reduction of support payments will, in general, reduce the intensity of agricultural production but could adversely affect highly prized habitats and landscape features unless this reduction is accompanied by the introduction of incentives for the maintenance of these areas.

A reduction in price support payments will, in general, reduce the intensity of agricultural land-use as the expected rates of return from agricultural investment fall. In addition, the uncertainty faced by farmers can be expected to increase and consequently they can be expected to diversify into a wider range of production activities. Both these factors should be beneficial to the environment. However, reducing output prices without the introduction of suitable compensation schemes will further impoverish many marginal farmers whose farms contain highly prized habitats and landscape features. This will lead to abandonment and neglect for some, and intensification for others. In neither case can it be guaranteed that ecological and scenic assets will be protected.

iii) Declining farm incomes will encourage larger farmers to disinvest in conservation measures, but make conservation payments more attractive.

Declining farm incomes may also force the larger farms to disinvest in on-farm conservation measures. Where land-use changes from cropping to grazing, or grazing to forestry some ecological advantages can be expected. In the long run, new habitats and new ecological systems will be created. Farms in financial difficulty may be bought by neighbouring landowners who can be expected to use the resultant larger farms differently. This should serve to increase the attractiveness of conservation and especially the acceptability of compensation payments. This environmentally desirable consequence would, however, not occur if conservation payments are reduced in parallel with price support.

iv) The introduction of financial assistance for environmentally enhancing agricultural practices will be necessary to prevent the reduction of output prices from having adverse effects on the quality of the environment in areas which have valuable scenic and/or ecological assets.

A reduction in price support would also cause land prices to fall, so the best farms from a conservation viewpoint would become less expensive and therefore more attractive to bona fide conservation bodies. But the very best properties may hold a "conservation premium", which might retain the high prices which specialist estates demand. In conclusion it would appear that financial assistance measures for guided land management and healthy agricultural practices should accompany the reduction of out put prices if the objective of harmonising agricultural production with the maintenance of the quality of the rural environment is to be realised in areas which have valuable scenic and/or ecological assets.

Chapter 6

THE IMPACT OF POLLUTION FROM OTHER SOURCES ON AGRICULTURE

T. Crocker (1)

INTRODUCTION

The modern industrial version of alchemy, which transforms otherwise harmless natural elements into a pervasive toxic burden, harms agriculture. Its air, soil and water pollution manifestations frequently reduce agricultural yields, increase the prices that consumers of agricultural products must pay, and alter the returns accruing to owners of agricultural inputs.

Many parts of any society regularly confront the problem of "optimal" ecosystem management. A farmer's choice of crops to grow, a householder's choice of yard plantings, and multiple-use management of a forest can all be cast as problems of ecosystem management. The ecosystem can then be seen as a process that produces desired outputs and services, and that may be facilitated by some human inputs (e.g. fertilisation) and hindered by others (e.g. atmospheric pollution).

Just as a manufacturing firm requires a description of its production process in order to allocate its resources efficiently, the ecosystem manager must have in mind a description of the process by which the outputs he desires are produced, and of what effects any factors that he may control have on these outputs. A model of the internal workings of the ecosystem is fundamental to economic analysis of management strategies. Questions about the optimal combination of inputs with which to produce one or more types of ecosystem output under assorted environmental conditions have therefore traditionally been attractive research topics for plant scientists and economists.

The first section of this summary explores the understanding that economists and plant scientists have acquired to date of pollution-induced vegetation (and, in passing, livestock) damages in agriculture and their economic consequences.

1. This summary of two consultant studies was prepared by Professor T. Crocker. These consultant studies were prepared by R.M. Adams, and T.D. Crocker and M. Linster.

The next section reviews plant science information about pollution-induced vegetation damages and discusses its policy relevance. Livestock damages are also briefly considered. A third section sets forth a general economic framework in which natural science studies of plant damage are a necessary component. Recent empirical studies of economic consequences take up the fourth section. The fifth section points to repairable faults in these studies and suggests problems that challenge the appropriateness of the analytical frameworks that support them. A discussion of the policy implications of the research being reviewed takes up the final section. Throughout, we adopt an expository treatment, avoiding detailed mathematical and statistical arguments, formulations and proofs. We try to provide enough information to allow the careful reader to assess the current state of knowledge, to evaluate new contributions, and to gain insight into the research efforts likely to offer the greatest improvements in current knowledge.

Natural Science Findings

Background

Table 6.1 lists the major air pollutants associated with agricultural impacts, their possible effects and major emission sources. Note that forests and fisheries are excluded from consideration, not strictly coming under agriculture.

Acid deposition occurs when sulfur dioxide and nitrogen oxides (NO_χ), emitted predominantly by sources that burn fossil fuels, are transported in the atmosphere and chemically transformed into acidifying substances before being deposited on the ground. Both precursors and transformation products can be transported hundreds or thousands of kilometers from the point of emission. In Europe, sulfur dioxide contributes about 70 per cent of acidity, nitrogen oxides 30 per cent. Photochemical oxidants are formed in the atmosphere by photochemical reactions involving anthropogenic emissions of NO_χ and volatile organic compounds (VOC's). VOC concentration is often the rate limiting factor; the major emission source is motor vehicles. Studies show that oxidant precursors and transformation products can be transported over great distances (several thousand kilometers). It is believed that photochemical oxidants play an important role in acid formation through the oxidation of sulfur and nitrogen oxides to acids.

In the last decade, SO_2 emissions have generally been falling or stabilizing in most OECD countries and are expected to continue to decrease. Since the early 1970s, NO_χ emissions in OECD have also shown a limited decline or at least stabilisation with the exception of southern Europe, and emissions of organic compounds are expected to continue to increase in most OECD countries in absence of additional control.

These trends suggest that neither acid deposition nor the presence of photochemical smog can be expected to diminish in the near future, and that only adoption of additional air pollution control measures would reduce the agricultural damage currently observed.

182

Table 6.1

SELECTED AIR POLLUTANTS, THEIR POSSIBLE IMPACTS ON AGRICULTURE, AND EMISSION SOURCES

Pollutant	Possible impacts		Emission sources
Acidic products (acid deposition)	Directly and indirectly (via soil changes) toxic to plants: foliar injury and inhibition of plant growth and development	Sulfur dioxide Nitrogen oxides	Power plants Fuel combustion (industrial, domestic, etc.) Industrial combustion processes Energy production Mobile (motor vehicle) sources
Photochemical oxidants (primarily ozone)	Foliar injury; inhibition plant growth and development	Nitrogen oxides Volatile organic compounds	(see above) Mobile (motor vehicle) sources Solvent use Natural
Sulfur dioxide (gas)	Foliar injury; inhibition plant growth; sulfur a plant macronutrient	Sulfur dioxide	(see above)
Nitrogen dioxide (gas) (see above)	Nitrogen dioxide like ozone; nitrogen oxide less well studied - considered less toxic; nitrogen a plant macronutrient		
Heavy metals (e.g. cadmium, lead, mercury)	Toxic to livestock and crops (esp. enzyme inhibition); magnification in food chain (cadmium); contamination of crops destined for animals and humans		Smelting industries Fossil fuel combustion Waste incineration Natural
Fluorine compounds and fluorides	Toxic to livestock (fluorosis in cattle)	Fluorine: Fluorides:	Ceramic industry Fertilizer production Chemicals industry Metal smelting industry Brick, tile, glass prod. ind. Coal combustion
Soot, smoke, dust Fine particulate matter	Direct contamination; reduced rate of photosynthesis		Road construction, Quarrying Open-cast mining Cement manufacturing Industrial processes Diesel engines Coal combustion

In the past 15 years, the percentage share of population linked up to a sewage treatment facility has steadily increased in OECD Member countries. This has led to progressively larger quantities of sludge which must then be treated and disposed of as effectively as possible.

Between 1970 and 1980 the volume of sludge produced rose by about 15 to 30 per cent depending on the country concerned, and current output in OECD countries is estimated at approximately 12 million tonnes of dry matter per year. In view of the trends in water quality standards and improved treatment systems, it may be assumed that the output will increase even further during the next ten years.

Sewage sludge can be of domestic or mixed (industrial plus domestic) origin depending on the type of network feeding the effluents to the treatment plant. Thirty to forty per cent of the sludge is currently used in agriculture or in similar fields (forestry, waste land reclamation, creation of urban green spaces, landscaping etc.), the remainder being sent to landfill, incinerated, or dumped at sea.

Vegetation Effects: Acid Deposition

Acid deposition possibly damages vegetation both directly, perhaps in combination with other air pollutants, and indirectly, through acid-induced soil and surface water changes. The direct effects have been studied mostly in forest trees; they involve primarily damage to the external protective surface, leaching of macro-nutrients, and disruption of germination. Little is known about the uptake and metabolic effects of acidifying substances. Whether they damage other types of vegetation in a similar way under field conditions remains to be demonstrated. An early study on crops under controlled environmental conditions suggested this may be so: simulated rainfalls of pH 3.6 damaged the protective surface, disrupted metabolism, growth, and reproductive processes, and decreased resistance to drought and other stresses particularly in soybeans, spinach, and alfalfa.

An attempt to quantify crop loss in the field began in the early 1980s in the United States under the National Acid Precipitation Assessment Program (NAPAP) which has as research goals to establish threshold doses that affect crops either adversely or beneficially and to develop mathematical models to estimate economic losses. Preliminary results showed no reduction in yield when potatoes, soybean and corn were exposed to simulated acid rain at pH ranges 4-4.5 or the current average ambient level for highly affected areas in the United States. The Amsoy variety of soybean did, however, show an 8 per cent decrease in yield between pH 4.2-5.6. The transitory foliar, necrotic lesions observed on soybean seedlings did not affect yield. Studies are planned to investigate the combined effects of wet and dry deposition and of oxidants.

Dry deposition includes gaseous SO_2, and gases readily enter plants through the stomata, openings in the epidermis which permit normal gas exchange. Some of the dissociation products in water are normal plant metabolites, which could correct deficiencies in soil sulfur, but an excess of intermediate oxidation products from the overload of metabolic pathways disturbs metabolic processes and inhibits plant growth and development. The SO_2 levels reached under this exposure situation are not likely to damage plants in the absence of other air pollutants, but rather to predispose plants to other stresses.

Indirect, long-term effects on plant productivity may result from acid-induced soil changes; possible consequences in the natural environment of a decrease in soil pH include an increased rate of leaching of important plant nutrients from the soil and decomposing vegetative matter (e.g. Magnesium and Cadmium), at a pH 5.0, an increased rate of mobilisation of potentially phytotoxic metals (e.g. Aluminum), and inhibition of the activity of pH-sensitive decomposer micro-organisms. In the last case, the pathway by which plant nutrients are cycled is blocked thereby reducing soil fertility. Soils in different regions vary in their resistance to acidification. Soil buffering capacity is determined primarily by type of bedrock and soil; base-poor, slowly-weathered rocks are most at risk from acidification, and are widely found in northern Europe, China, and eastern North America. Acid deposition is likely one of the many causes of soil acidification.

Soil acidification also occurs naturally and by land management practices, e.g. the use of fertiliser containing ammonium nitrogen. Agricultural soil is probably the least affected by acid deposition because it is regularly fertilised and limed. In Sweden, fertiliser containing ammonium nitrogen contributes more to acidification of agricultural soil, estimated at about 20 per cent, than acid deposition, estimated at 10 per cent. The problem is that not enough lime is being added to counter soil acidification, and there is an accumulated lime demand. An estimated 40 out of 200 kg of lime per hectare per year is required to neutralize fertilisers, and 10-30 kg to offset acid deposition. Where soils are low in nitrogen or sulfur, some beneficial impact on plants may even result from limited acid deposition.

No information is available on the direct and indirect impact of acid deposition on orchard trees. Since they are exposed for several years to acid deposition in combination with gaseous pollutants and other stresses (disease, extremes in climate), it is tempting to speculate that the impact is greater than on annual crops, depending on land management practices of course, and is most likely manifested by reduced growth and increased susceptibility to abiotic and biotic stresses. In addition, the impact on rangelands needs to be studied.

In summary, the nature, intensity, and extent of the acid deposition impact on agriculture remains largely unknown. Direct adverse effects on field-grown crops may be limited to a few sensitive species. The main indirect adverse effect may be the increase in lime demand in regions with soil susceptible to acidification. The need for more research must be emphasised: cause-effect and, as a second step, dose-response relationships have still to be clarified at present levels of acid deposition.

Vegetation Effects: Photochemical Oxidants

The principal component of photochemical oxidants is ozone (O_3). It is often used as an indicator of the degree of oxidant pollution, and its adverse effect on plants has been studied the most. Other potentially phytotoxic components of photochemical oxidants are nitrogen oxides and peroxyacetyl nitrates (PAN). Plants are generally less sensitive to nitrogen dioxide and nitrogen oxide than to O_3; NO_2 in a mixture with O_3 and SO_2 presents a greater risk of plant damage because of synergistic effects, whereas NO, although less well studied, is considered less toxic. PAN is most phytotoxic, but levels are too low in most regions to pose a significant risk to vegetation. The emphasis here, therefore, is on the role of ozone in crop damage.

The nature of ozone damage is being elucidated in controlled chamber and greenhouse studies, although the exact mechanism of toxicity remains unknown. Ozone, like other gaseous pollutants, enters into the leaf by passive diffusion through open stomata and is then dissolved or hydrated in the extracellular water. The opening of stomata, which is controlled genetically and modified by environmental factors, such as wind speed, temperature, and relative humidity, and solubility of the gas in extracellular water influences the rate of uptake. Ozone breakdown products include free radicals which, because of their highly reactive nature, have been hypothesized to exert a toxic effect on the cell wall or plasma membrane. Consequent biochemical and physiological changes include disruption of cell membrane integrity and of metabolic processes such as photosynthesis.

Damage occurs when the plant is unable to metabolize the ozone or its metabolites, or to repair the negative effect. The effects may be manifested as foliar or visible injury, reduced plant growth and crop yield, altered product quality, increased evapo-transpiration and changes in susceptibility to abiotic (drought or frost) or biotic (insect or pathogen) stresses. Foliar injury is often associated with short-term, high-level exposure to ozone (acute response) while reduced crop yield follows long-term, low-level exposure (chronic response). Either type of damage may occur in the absence of the other. Some earlier studies of ozone-induced plant damage measured visible damage only; for food and forage crops, however, reduced yield is generally considered to be the better criterion of damage unless the appearance of leaves is important in determining the plant's value (e.g. spinach, lettuce, tobacco). Whether reduction in yield is an outcome may depend on harvest date for some plants. It has been found, for example, that barley grew and developed slower in ambient air compared with filtered air (50 per cent less ozone), but by harvest time the yields under the two exposure conditions were not significantly different.

The nature of combined pollutant effects is less well known than that of single pollutants and is currently an important topic of research. Briefly, the following can be said about interactive effects based on other reviews of the literature: the mixture of O_3 and SO_2 is synergistic and individual concentrations which do not inhibit growth if combined show additive or antagonistic effects; individual concentrations of NO_2 and O_3 which are non-toxic induce visible damage when mixed; the three gaseous pollutants combined may have synergistic or additive effects. The interaction between gaseous pollutants, especially O_3 and SO_2, and acid deposition is also under investigation.

The intensity of effect (and even the nature of the effect) depends on O_3 concentration, exposure duration and frequency, intervals between exposure, time of day and year, prevailing environmental conditions, etc. None of the statistics available adequately characterise the dynamic nature of plant exposure, and a variety of measurements exist in the literature. Overall mean concentration during daylight hours and frequency of episodes above a certain threshold of concentration are considered to be among the more acceptable measurements.

186

Table 6.2 displays the lowest ozone exposures by exposure durations and qualitative plant sensitivities that plants can tolerate. Sensitive plants are likely to be harmed if ozone exposure exceeds 100 ug/m³ for 4 hours to 300 ug/m³ for 0.5 hours; less sensitive plants can withstand ozone concentrations about three times higher for the same exposure duration.

Table 6.2

PROPOSED MAXIMUM ACCEPTABLE OZONE CONCENTRATIONS
IN ug/m³ FOR THE PROTECTION OF VEGETATION

Exposure (duration, hours)	Resistance Level		
	Sensitive	Intermediate	Less Sensitive
0.5	300	500	1 000
1.0	150	350	500
2.0	120	250	400
4.0	100	200	350

Source: Guderion (1984)

Grasses and cereals, which include the most economically important crops in OECD countries are, as a group, intermediate in sensitivity.

Table 6.3 gives examples of yield reduction as a function of O_3 concentration and exposure duration for several plants tested under various experimental conditions. The data indicate that, in general, exposure to ozone at 200 ug/m³ for a few hours per day for several days or weeks reduces significantly crop yield (in most cases by at least 20 per cent). This table also shows the variety of plants susceptible to ozone damage. For comparison, ambient daily ozone concentrations in rural and relatively remote areas were 200 ug/m³ (with highs of 340 ug/m³ in Canada to lows to 100 ug/m³ in Japan) in the latter half of the 1970s and early 1980s. As can be seen, the above response data suggest that visible injury and reduced yield are already occurring in sensitive plants at ambient ozone concentrations; however, the plant data were obtained in North America, and whether they can be extrapolated to the conditions of other regions in the world remains an open question.

Because of limited or unreliable results quantifying crop response to air pollution and the environment, an ambitious research project was undertaken by the National Crop Loss Association Network (NCLAN, set up by the United States Environmental Protection Agency in 1980) to generate dose-response data for major agricultural crops under field conditions in several production areas of the United States. Plants were exposed throughout the growing season to a range of ozone concentrations for 7 hours/day in

Table 6.3

OZONE CONCENTRATIONS AT WHICH SIGNIFICANT YIELD LOSSES HAVE BEEN NOTED FOR A
VARIETY OF PLANT SPECIES EXPOSED TO O_3 UNDER VARIOUS EXPERIMENTAL CONDITIONS

Plant species	O_3 concentration[1] (g/m^3)	Exposure duration	Yield reduction % of control	Year of study
Alfalfa	200	7 h/d, 70 d	51, top dry wt	1977
Alfalfa	200	2 h/d, 21 d	16, top dry wt	1975
Pasture grass	180	4 h/d, 5 d/wk, 5 wk	20, top dry wt	1980
Ladino clover	200	6 h/d, 5 d	20, shoot dry wt	1982
Soybean	200	6 h/d, 133 d	55, seed wt/plant	1974
Sweet corn	200	6 h/d, 64 d	45, seed wt/plant	1972
Sweet corn	400	3 h/d, 3 d/wk, 8wk	13, ear fresh wt	1973
Wheat	400	4 h/d, 7 d	30, seed yield	1974
Radish	500	3 h	33, root dry wt	1974
Beet	400	2 h/d, 38 d	40, storage root dry wt	1973
Potato	400	3 h/d, every 2 wk, 120 d	25, tuber wt	1980
Pepper	240	3 h/d, 3 d/wk, 11 wk	19, fruit dry wt	1979
Cotton	500	6 h/d, 2 d/wk, 13 wk	62, fiber dry wt	1979
Carnation	100-180	24 h/d, 12 d	74, no. of flower buds	1968
Coleus	400	2h	20, flower no.	1972
Begonia	500	4 h/d, once every 6 d for a total of 4 times	55, flower wt	1979
Ponderosa pine	200	6 h/d, 126 d	21, stem dry wet	1977
Western white pine	200	6 h/d, 126 d	9, stem dry wet	1977
Loblolly pine	100	6 h/d, 28 d	18, height growth	1977
Pitch pine	200	6 h/d, 28 d	13, height growth	1977
Poplar	80	12 h/d, 5 mo	+1333, leaf abscission	1977
Hybrid poplar	300	12 h/d, 102 d	58, height growth	1981
Hybrid poplar	300	8 h/d, 5 d/wk, 6 wk	50, shoot dry wt	1981
Red maple	500	8 h/d, 6 wk	37, height growth	1979
American sycamore	100	6 h/d, 28 d	9, height growth	1982
Sweetgum	200	6 h/d, 28 d	29, height growth	1982
White ash	300	6 h/d, 28 d	17, total dry wt	1982
Green ash	200	6 h/d, 28 d	24, height growth	1982
Willow oak	300	6 h/d, 28 d	19, height growth	1982
Sugar maple	300	6 h/d, 28 d	12, height growth	1982

1. Data originally given in ppm; conversion factor used 1 ppb = 2 g/m^3
   Data represents the lowest concentration reported that significantly reduced yield; this
   concentration was frequently the lowest tested.

Source: Tingey (1984).

daylight, when plants are most sensitive, photosynthetic activity is greatest, stomata are open and peak ozone concentrations occur. Eight species (corn, soybean, kidney bean, head lettuce, peanut, spinach, turnip, and wheat) and 18 cultivars were initially tested and yield response models were developed for each cultivar (control: filtered air with 50 ug/m^3 O$_3$). For several species, 7-hour seasonal mean O$_3$ concentrations exceeding 80 to 100 ug/m^3 (data originally given in ppb; 1 ppb = 2 ug/m^3) were predicted to reduce yields by 10 per cent for soybean, 14 to 17 per cent for peanuts, 53 to 56 per cent for head lettuce, and 2 per cent for red kidney beans.

Results from a limited number of open-top field chamber studies performed in Denmark, Great Britain and Sweden also showed both visible damage and reduced yield in sensitive varieties of crops, such as spinach, peas, beans and potatoes, occurring at ambient levels recorded during the study or at levels often reached in these countries. It is quite likely that effects had been under-estimated in the past, for example by attributing possible O$_3$-related damages to another unknown stress, as was the case with pea Pisum Sativum crops in the United Kingdom.

In summary, ozone's phytotoxic properties are well established, but interactive effects of combined air pollutants need more study, and until then, the intensity of plant response to field pollution mixtures cannot be quantified. The general consensus emerging in many OECD countries is that ambient ozone levels are having an adverse impact on agricultural crops and that in the past its effects have been under-estimated. Several types of data support this: results from experiments conducted under controlled environmental conditions indicate that 50 to 200 ug/m^3, levels often reached in rural areas, cause visible injury and reduce yield in several plant species; field investigations using bio-indicator plants and open-top field chambers also showed plant damage at ambient O$_3$ concentrations (generally around 100 ug/m^3 -- 7-hour mean).

Vegetation Effects: Sewage sludge

Trace metals such as lead, cadmium, zinc and copper can inhibit plant growth. Any toxic effect they have depends not only on soil amounts but above all on their concentration in the soil solution. The metals present in the liquid phase are those which can be directly taken up by the plant roots. The effect of soil characteristics (pH, organic matter content, etc.) is paramount in this respect. The more acid a soil, the less it retains heavy metals, which enter the liquid phase and thereby become more available to plants.

The element presenting the greatest toxicity hazard is cadmium, owing to its high mobility through the food chain. It is known to be highly available for absorption by plants, unlike other elements, such as lead, which are difficult to absorb.

There is no doubt that the spreading of sludge increases heavy metal concentration in the soil solution. The combined supply of organic matter may, however, limit (at least initially) the transfer of heavy metals to the liquid phase. In general, heavy metals from inorganic sources are thought to be more available to plants than those from organic sources.

Comparative studies of yields obtained following the application of sludge with low trace metal concentrations and of sludge with high concentrations have shown a slight decrease in crop yields for highly contaminated sludge. A French study on crops having received massive amounts of sludge polluted by cadmium and nickel showed yield losses of 20 per cent for maize crops. The lower yields were accompanied by yellowing of the older leaves, increased heavy metal concentration in leaves and a general decrease in the phosphorus content. On the other hand, nickel chiefly accumulated in seed. For other types of crops such as lettuce, yields were not affected in spite of a higher heavy metal concentration in the leaves. The effects observed for average doses were fairly similar to those observed with massive ones. It therefore seems that plants growing on soil with a high trace metal content rapidly reach a maximum level beyond which further sludge application has little incremental effect.

The accumulation of trace metals in the top soil not only creates a hazard for plants but also for grazing animals, which may ingest the metals directly. Entry into the food chain therefore occurs through both crops and contaminated animals. The most sensitive crops to the accumulation of heavy metals are fruit and vegetables.

The presence of pathogens in sludge is chiefly a hazard to grazing animals and has no effect on plant growth and crop yield. Plants are regarded above all as "the vectors" of pathogens. A comparison between the infection rate among cattle grazing on land treated with sludge and on untreated land has revealed significant differences. In the United Kingdom however, research on this aspect has not revealed any evidence of a direct link between sludge and diseases. The problem mostly arises in areas where land available for sludge spreading is limited, the animal population density high and the pathogen content, especially that of Salmonellas, of sludge is high (use of untreated raw sludge). Apart from Salmonellas, no other species of bacteria seems to be transmitted by sludge. As for parasites, it has been shown that sludge can act as a vector for worms such as thread and tape worms. For other species, the role of sludge is much less certain and would require additional research and more systematic epidemiological monitoring. Very little work has also been done concerning viruses.

The least known pollution from sludge concerns organic compounds, especially since there have not been any clear indications so far on any harmful effects on plants, soil or man following agricultural spreading.

Vegetation Effects: The Value of Information

A basic science, piece-by-piece approach to the study of how pollution affects vegetation makes truly awesome the task of covering and synthesizing the feasible or even the technically efficient input combinations. In agriculture alone, environmental and edaphic factors create thousands of different types of inputs. Each input type is, in turn, embodied in one or more production processes or crops, which may appear in a variety of cultivars and which can be put to many distinct uses. Moreover, environmental cofactors, such as moisture and temperature, may act in concert with pollution to alter its impact. Natural science studies of how pollution affects plants have neither received nor provided much guidance about which of these many input combination have any economic significance.

One key problem, then, is to reconcile the natural science and the economic approaches to plant damages by developing criteria for deciding how much plant science information to generate and retain in any particular pollution problem setting. When one of the several objectives of plant science pollution damage studies is to provide information useful for estimating economic consequences, a basic question is whether more or less natural science detail will alter the economic estimates in a non-trivial way. Several studies have now demonstrated that there are some important classes of pollution impacts upon agriculture in which rather imprecise yield-response information is quite adequate for distinguishing among the economic consequences of alternative ambient pollution levels.

In a Bayesian inquiry into the robustness of economic estimates of crop damages from ozone, one study showed that the policy value of additional plant science yield response information declines rapidly. In an entirely different setting, it was demonstrated that the value of information to Israeli growers about potato yield responses to soil salinity was maximized with as few as 15 and not more than 27 observations. While studying the economic benefits of ambient ozone control for corn, soybeans and wheat grown in the US corn belt, it was found that changes in key natural science parameters had to be substantial if they were to translate into major changes in benefit estimates.

The latter study, however, adds two strong qualifications to this conclusion. First, they note that some researchers hypothesize that climatic and soil factors are important in modelling the real-world effect of ozone on crop yields. When response functions reflecting moisture stress-ozone interactions were included in their economic analysis, the changes in benefit estimates were as great as 50 per cent.

This study also warns that it is only the latest in a succession of studies showing that yield response and economic benefit estimates are sensitive to the choice of form for the yield response function. The benefit estimates differed by as much as 60 per cent according to whether a linear or a nonlinear yield response function was employed. This result is similar to that previously found in previous studies between linear and quadratic forms.

Whatever the correct form of the yield response function, most efforts to fit observed yields statistically to air pollution exposures have employed classical least squares techniques to account for random variation. One argument is that the possible non-normality of yield response data justifies the use of maximum likelihood estimators rather than classical least squares. With data on the yield responses of soybeans to ozone, it is shown that maximum likelihood estimates for nonlinear distribution are about 25 per cent greater than the least squares estimates for the same distribution and similar data.

The Economic Assessment Problem

Assessments of the economic consequences of pollution upon agriculture try to estimate differences in income equivalents, defined as differences in the sums of buyer "surpluses" and seller "quasi-rents" over two or more policy relevant pollution levels. Buyer (usually called consumer) surplus portrays the difference between the maximum a representative individual would be willing to commit himself to pay for a commodity unit and what he in fact has to pay. Similarly, seller (usually called producer) quasi-rent is the difference between what a commodity owner receives for supplying a commodity unit and the minimum he must receive in order to be willing to commit to that supply. Note that both definitions are ex ante and are therefore capable of capturing attitudes toward risk. The sum of seller quasi-rent and buyer surplus is thus a measure of the prospective net benefits from the production and the consumption of a commodity. The observable unit prices of other commodities that could provide him equal satisfaction set an upper bound to the buyer's maximum willingness-to-pay; the earnings his resources could obtain in other activities set a lower bound on the minimum reward the seller must receive. Maximum willingness-to-pay represents demand; the minimum necessary reward defines supply.

A thorough assessment of pollution control benefits, the gainers' income equivalents, requires three kinds of information: 1) the differential changes that pollution control causes in each person's production and consumption opportunities; 2) the responses of input and output market prices to these changes; and 3) the input and output changes that those affected can make to minimize losses or maximize gains from changes in production and consumption opportunities and in the prices of these opportunities. Natural science studies of biological dose-response functions are the primary source of information for the first requirement. Evaluation of the latter two requirements represents the economics portion of any benefits assessment exercise. If pollution control causes substantial changes in outputs, price changes can occur which, in turn, lead to further market-induced output changes. Moreover, even if prices are constant, natural science information will still fail to provide accurate indications of output changes when individuals can alter production practices and the types of outputs produced. Thus, accurate information on the economic consequences for agriculture of pollution can be achieved only if the reciprocal relations between physical and biological changes and the responses of individuals and institutions are explicitly recognised.

Empirical Assessments

Tables 6.4 and 6.5 respectively portray the conditions specified and the empirical results obtained for selected studies of the economic consequences of air pollution for North American, particularly US agriculture. Caution should be exercised in trying to compare numerical estimates across studies since crops, response information and assumed conditions differ considerably. In no study do estimated control benefits exceed 5 per cent of total crop value. However, though numerical estimates and the conditions under which they were derived may differ, these divergent studies exhibit common qualitative patterns of behavioral responses and sensitivities to imposed conditions.

Increasing air pollution causes losses in the total economic surplus from the production and consumption of agricultural outputs to increase at an increasing rate. Most of the relevant studies deal solely with ozone. Only a few studies deal with more than two ambient pollution levels, thus making it impossible to use them to evaluate the rate of change in economic surplus with respect to changes in pollution levels. No thoroughly consistent pattern emerges as to the absolute magnitudes of this rate of change in total surplus.

Table 6.4

SELECTED RECENT ECONOMIC STUDIES OF AIR POLLUTION IMPACTS ON UNITED STATES AGRICULTURE

Study	Pollutants	Ambient Concentrations	Completeness				Crops	Results (1980 US Dollars)		
			Price Changes	Crop Substitutions	Input Substitutions	Quality Changes		Consumer Benefits	Producer Benefits	Total Benefits
Stanford Research Institute (1981)	Ozone	1) Universal reduction to 80 ppb.	No	No	No	No	Corn, soybeans, alfalfa, and 13 other annual crops	None	$1800x10^6	$1800x10^6
	SO$_2$	2) Universal reduction to 260 ng/m3						None	23x10^6	23x10^6
Shriner, et al. (1982)	Ozone	3) Universal reduction to 25 ppb.	No	No	No	No	Corn, soybeans, wheat, peanuts	None	3000x10^6	3000x10^6
Kopp, et al. (1985)	Ozone	4) Universal reduction from 53 ppb. to 40 ppb.	Yes	Yes	Yes	No	Corn, soybeans, wheat, cotton, peanuts	Not present	Not present	1300x10^6
Adams and Crocker (1984)	Ozone	4) Universal reduction from 53 ppb. to 40 ppb.	Yes	Yes	Yes	No	Corn, soybeans, cotton	Not present	Not present	2200x10^6
Adams, Crocker, and Katz (1984)	Ozone	4) Universal reduction from 48 ppb. to 40 ppb.	Yes	No	No	No	Corn, soybeans, cotton, wheat	Not present	Not present	2400x10^6
Adams, Hamilton, and McCarl (1986)	Ozone	4) Universal reduction from 53 ppb. to 40 ppb.	Yes	Yes	Yes	No	Corn, soybeans, cotton, wheat, barley	1160x10^6	550x10^6	1700x10^6
Shortle, Dunn, and Phillips (1986)	Ozone	4) Universal reduction from 53 ppb. to 40 ppb.	Yes	No	No	Yes	Soybeans	880x10^6	-90x10^6	790x10^6
Adams, Callaway, and McCarl (1986)	Wet acid deposition	Universal decrease from 4.8 to 4.5	Yes	Yes	Yes	No	Soybeans	-172x10^6	30x10^6	-142x10^6

1) Averaging time of one hour; not to be exceeded more than once a year.
2) Averaging time of 24 hours; not to be exceeded more than once a year.
3) Annual geometric mean.
4) Seven-hour growing season geometric mean. Given a log-normal distribution of air pollution events, a 7-hour seasonal ozone level of 40 ppb is equal to an hourly standard of 80 ppb, not to be exceeded more than once a year [(Heck, et al. (1982)].
e) Annual arithmetic mean of rainfall pH.

Table 6.5

SELECTED REGIONAL STUDIES OF AIR POLLUTION IMPACTS ON AGRICULTURE

STUDY	Pollutants	Ambient Concentrations	Completeness				Crops	Place	Total Benefits (1980 US Dollars)		
			Price Changes	Crop Substitutions	Input Substitutions	Quality Changes			Consumer Benefits	Producer Benefits	Total Benefits
Forster (1984)	Wet acid deposition	5) Universal increase to 5.6 pH	No	No	No	No	All	Eastern Canada	Not present	75x106	105x106
Crocker and Regans (1985)	Wet acid deposition	5) Universal increase to 5.6 pH	No	No	No	No	All	Eastern US	Not present	130x106	130x106
Page, et al. (1982)	Ozone	1) Universal reduction to 80 ppb	No	No	No	No	Corn, soybeans and wheat	Ohio River Basin, US	Not present	Not present	400x106
Adams, Crocker and Thanavibulchai	Ozone	1) Universal reduction to 80 ppb	Yes	Yes	Yes	No	Corn, sugar beets, and 12 annual vegetables	Southern California	14x106	51x106	65x106
Adams and McCarl (1985)	Ozone	1) Universal reduction to 80 ppb.	Yes	Yes	Yes	No	Corn, soybeans, and wheat	Corn Belt, US	2079x106	-1411x106	668x106
Howitt, Gossard and Adams (1984)	Ozone	4) Universal reduction to 40 ppb	Yes	Yes	Yes	No	38 crops	California	17x106	28x106	45x106
Mjelde, et al. (1984)	Ozone	4) 10% increase from 46.5 ppb	No	Yes	Yes	No	Corn and Soybeans	Illinois	None	-226x106	-226x106

1) Averaging time of one hour; not to be exceeded more than once a year.
2) Averaging time of 24 hours; not to be exceeded more than once a year.
3) Annual geometric mean.
4) Seven-hour growing season geometric mean. Given a log-normal distribution of air pollution events, a 7-hour seasonal ozone level of 40 ppb is equal to an hourly standard of 80 ppb, not to be exceeded more than once a year [(Heck, et al. (1982)].
5) Universal increase to an hourly standard of 80 ppb, not to be exceeded more than once a year [(Heck, et al. (1982)].

Growers can gain from increases in air pollution. It was found that substantial increases in acid deposition enhance the quasi-rents of soybean growers. These growers benefit similarly from ozone increases, with the same result for corn, soybean and wheat growers, considered as a package. A sufficient condition for these findings is that the pollution-induced reduction in output quantity be less than the percentage increase in market price that the supply reduction causes. All other studies conclude that growers lose from increased air pollution. However, those studies which disregard market price changes and which therefore guarantee producer losses with increased pollution cannot be taken seriously as evidence supporting the hypothesis of producer losses.

Losses in buyer or consumer surpluses are a very significant portion of the total losses that air pollution causes agriculture. Of the studies that accounted for air pollution-induced changes in the prices of agricultural outputs, consumer surplus losses as a percentage of total losses range from a low of 22 per cent to a high of 100 per cent. Given that producers can sometimes benefit from air pollution increases, methods which disregard consumer impacts can, it seems, understate total losses in surplus and grossly misstate the distribution of these welfare effects.

For air pollution increases, percentage losses in total economic surplus are no greater than and nearly always less than the percentage changes in biological yields which triggered these losses; for air pollution decreases, percentage gains in producer surplus are no less than and nearly always greater than the percentage changes in biological yields which triggered these gains. Every study which accounted for price effects or crop and input substitutions obtained these results. The results occur because economic agents tend to substitute toward those crops experiencing advantageous relative price and production changes with a pollution reduction, and away from those crops suffering disadvantageous changes with a pollution increase. Substitutions enhance the advantageous changes for the agent, and attenuate the disadvantageous changes.

Air pollution increases can cause growers of relatively pollution-tolerant crops to suffer production declines greater than those triggered by the pollution increase. This phenomenon is explicitly noted in research in which growers of pollution-tolerant crops like broccoli, cantaloupe, carrots and sugar beets were also estimated to experience production yield declines as air pollution increased. Furthermore, several studies of the corn-soybean-wheat economy in the United States midwest have produced the same finding. This counter-intuitive result occurs because of a "crowding out" phenomenon. If, after a pollution increase, the pollution-intolerant crops continue to have higher unit values than the pollution-tolerant crops, and if inputs such as land and water are scarce, growers of the pollution-intolerant crops will substitute land, water and similar inputs for air quality. The more profitable affected crops now require more land and water to produce the market equilibrium quantities.

Changes in air pollution affect both the productivity and aggregate demand for factors of production. It is estimated that a 10 per cent increase in ozone results in a 4 per cent decline in the demand for variable inputs such as labor, water and fertiliser. One study found that the ambient ozone conditions prevailing in southern California during the mid-1970s reduced the economic productivity of selected agricultural workers by an arithmetic mean of 2.2 per cent, with a range from zero to 7.4 per cent Research has shown that increased ozone increases the demand for land inputs.

Changes in air pollution have differential effects on the comparative advantage of agricultural production regions. Studies indicate that reductions in air pollution resulted in gains to producers in areas simultaneously characterized by high ambient pollution levels and crop mixes crops dominated by air pollution sensitive crops. Such gains were frequently in the form of expanded crop acreage. Conversely, areas with the opposite set of characteristics experienced reductions in acreages of some crops and hence a loss of "market share" for these crops. While these analyses all found net total gains to society from reductions in air pollution, some subregions gained at others' expense.

Air pollution has economic transboundary effects that alter international trade flows with attendant gains and losses to exporters and importers. In the United States, reductions in air pollution led to increases in production of agricultural commodities, with much of this increase moving into exports. The net effect is that foreign consumers of these commodities capture most of the consumer gains. Indeed some researchers have estimated that of the total consumer gains from ambient ozone reductions, 60 per cent accrued to non-United States consumers. Conversely, increases in air pollution imply a reduction in the welfare of importing countries. As a result, national environmental policies can readily have economic transboundary implications even in the absence of a pollution transboundary phenomenon.

The above commonalities emerge from the substantial number of studies about air pollution effects upon agriculture. Assessment studies about pollution impacts other than air are rare. It is highly likely, however, that most of the aforementioned qualitative observations apply equally well to other pollutants and to other settings, including sewage sludge.

Gaps and Challenges

Natural Science Gaps

Knowledge about the dynamic (sequential) responses of vegetation to pollution stresses is, at best, superficial. The dynamics of these responses have implications for both annual and perennial cropping systems. For example, for annuals, we know that the pool of genetic material available to growers tends to change over time. Hence, analyses performed today may not be relevant to the gene pool that will be available in the future. More importantly, however, there are large categories of vegetation for which we know little or nothing about responses over time. Perennial plants such as orchard species are one clear example where current knowledge is limited mostly to guesses as to long-range consequences.

A related natural science issue relates to potential interactions among pollutants and between pollutants and other environmental stresses. The existing response literature consists mostly of studies on the effects of individual pollutants on individual crops. This crop by crop, pollutant by pollutant approach to crop response is a severe abstraction from real world conditions. It cannot be a tractable approach if economic assessments are to address pollution impacts of broad geographical scope. At least some future plant science research must focus on acquiring a structured and robust understanding of the fundamental processes underlying crop responses to pollution; that is, timely policy responses to pollution problems of large geographical scope require the development of generic plant response information. Simultaneously, more precise information about actual growing season daylight pollution levels at agricultural sites must be acquired.

Economic Analysis Gaps

Though existing studies address a wide range of the economic consequences of pollution, opportunities remain to improve the dimensionality of the models. One such opportunity deals with the grower uncertainties induced by prospective changes in pollution types and levels. Specifically, if changes in pollution events increase the natural variability of crop yields, the risk averse grower confronts additional costs; he will expend resources preparing for world states no more than one of which will be realized at any given time. No assessment study to date has explicitly incorporated risk or the income variability associated with pollution events into its analysis.

The ruling economic assessments universally assume that agricultural outputs are sold in long-term, perfectly competitive market settings where government programmes have no direct role. If one accepts the validity of the social goals which various government farm support programmes express, then correct assessments of the net economic consequences of pollution changes makes obligatory the inclusion in the analysis of the programme provisions. From the perspective of the programmes, pollution reductions which lead to supply increases are costly, if only because they increase government support costs. In addition, from a strictly market perspective, they expand the agricultural use of inputs such that their social costs exceed the value of their contribution to production.

Another issue now receiving little attention in the literature concerns potential changes in crop demand associated with pollution changes. With a single exception, all analyses focus solely upon changes in supply. However, pollution may change the qualities of some commodities. For example, ambient ozone can affect the protein content of soybeans and sewage sludge spreading can increase the trace metal burden of plants. These potential demand consequences have yet to be addressed.

Challenges

Some pollution effects on agriculture are a worldwide phenomenon and may require a different policy perspective. For example, the current scientific and policy discussions concerning climate change involve a complex set of pollutants/climate interactions that, taken together, have true global implications. Analysis and resolution of a global pollutant issue creates a set of new challenges for researchers and policymakers. The first such challenge pertains to the different orderings of economic processes, markets and institutions across countries. This diversity requires an assessment framework that both exploits the common relationships between economic orderings while adequately representing the relevant individual characteristics of each country or region, including planned economies and subsistence, village-oriented agriculture. The second challenge relates to a fundamental lack of biological data from which to predict crop responses across matrices of pollutant, site and time combinations. Innovative procedures are required to determine the "transferability" of response parameters within this matrix. A crucial starting point in economic analysis is the correct representation of the choice problem facing the various decision-makers in each economic setting.

However desirable it may be to develop complete economic representations of subsistence and other behavioral models, in the short term tractable research on the global consequences of pollution may be limited to gross evaluations of changes in world trade flows in agricultural commodities. Since world trade in agricultural commodities tends to respond to market (price) signals, the competitive models used in some existing trade flow studies may be appropriate. Specifically, a general equilibrium framework that encompasses agricultural production and consumption in each country, and that then captures the net effect of each country on world trade is required. In practice, existing econometrically-based world trade models of agriculture, such as those of the US Department of Agriculture, may be one means of evaluating aggregate effects of air pollution or other environmental changes.

Wherever they have been done, prospective and retrospective assessments of pollution impacts upon agriculture provide convincing evidence of substantial economic effects having plausible policy relevance. Nevertheless the weight of the evidence cannot be interpreted as a call for a proliferation of biological and economic studies unique to each site, time, and food and fiber allocation system. Research funds and talents in any country are too valuable to permit this. The development of generic dose-response relations will greatly enhance model robustness, where robustness can be defined as the domain of circumstances where the model can be applied without undergoing structural revisions. Development of generic models will take a long time, however. An alternative with a potentially immediate payoff is to assume that a grand model exists which carries across nearly all sites, times and allocation systems. One can then systematically exploit the exchangeability (transferability) concepts of Bayesian statistics. These concepts test the existence of a common structure which generates random samples drawn from a number of distinct groups. They allow one to draw systematic and communicable inferences about the magnitude of a parameter in one group from the observations on all groups. Even though transferability may be imperfect, numerical measures of the errors resulting from a transfer are generated. Policymakers and researchers can then decide whether this error is tolerable for the decision problem in question.

Some Policy Implications

Abundant empirical evidence exists that air pollution, especially its photochemical oxidant form, has substantial economic efficiency consequences involving regional and time-specific losses of up to 5 per cent of the economic surpluses associated with the production of annual crops. Percentage differences in distributional (equity) impacts across producers and consumers, importers and exporters, production regions, and individual crops are much greater. Indeed, the adjustment problems that these equity impacts raise for local and regional economic structures could be a major policy issue. Even in instances where agricultural impacts are ambiguous, such as acid deposition, the evidence is strong that particular groups and regions bear disproportionate shares of the risks that these potential impacts pose. A concentration upon levels of acceptable risk of pollution impacts upon agriculture should not be allowed to obscure the manner in which this risk is allocated.

Most of the evidence reviewed here about the levels of risk from prevailing ambient concentrations of photochemical oxidants cause producers and consumers of annual crops is derived from conceptually sound economic models and the best available plant science and aerometric data. There nevertheless remain important knowledge gaps, including inadequate information on response dynamics, environmental interactions, and the long-term and possibly irreversible effects of both gaseous and other pollutants on perennials. With the potentially outstanding exception of impacts upon perennials, it is unlikely, however, that inclusion of these possible improvements in future assessments would greatly alter estimated overall benefits of reducing pollution levels. The contribution of additional yield response information to the ability to discriminate among the aggregate economic impacts of alternative ambient pollution levels declines quite rapidly. Bioeconomic sensitivity analyses suggest that changes in key natural science parameters must generally be substantial in order to translate into major changes in aggregate economic estimates. It is improbable that new natural science findings concerning functional forms, dose measures, or interaction effects for agriculture would exceed the magnitude of yield changes already addressed in some of the sensitivity analyses. There also are likely to be countervailing responses that tend to offset errors, e.g. longer exposure periods may predict greater yield losses, but a combination of ozone and water stress tends to dampen yield losses attributed to ozone.

If more exact natural science information is to be of substantial policy value, it will probably be due to its contribution to better understanding of the manner in which the consequences of pollution impacts are distributed across groups and regions. The evidence reviewed makes it appear that production and consumption patterns, and thus distributional consequences, are often very sensitive to rather small changes in ambient pollution. At this time, it is not known whether estimates of the human behavioral responses which are the sources of this sensitivity would be greatly altered by more exact natural science information.

The policy relevance of improved assurances about the sensitivities of production and consumption patterns to variations in pollution levels is perhaps most interestingly demonstrated by the consistent finding that agriculturists substitute other inputs, especially land, for poor environmental quality; that is, reduced environmental quality causes more land and inputs such as fertiliser and pesticides to be used. At this juncture, only the act of substitution has been grasped. Its magnitude and the conditions that cause this magnitude to vary are presently little understood, although the finding that producer losses often increase exponentially implies that the worth of substitution possibilities progressively decline with increasing pollution.

Good natural science information is a prerequisite to a precise mapping of the substitution possibilities. Its importance becomes apparent when one recognizes that these land, fertilisers and pesticide substitutions could be responsible for a good deal of the pollution that agricultural practices cause. Policymakers might then confront the paradox that the pollution from other sources which affects agriculture encourages agricultural practices which produce pollution. Conversely, a reduction in pollution from other sources could reduce the pollution that originates in agricultural practices. A thorough understanding of whether this quite plausible conjecture is, in fact, true will require an as yet mostly unavailable detailed natural science and economic understanding of influential substitution possibilities and agriculturist behavioral responses.

WHERE TO OBTAIN OECD PUBLICATIONS
OÙ OBTENIR LES PUBLICATIONS DE L'OCDE

ARGENTINA - ARGENTINE
Carlos Hirsch S.R.L.,
Florida 165, 4º Piso,
(Galeria Guemes) 1333 Buenos Aires
Tel. 33.1787.2391 y 30.7122

AUSTRALIA - AUSTRALIE
D.A. Book (Aust.) Pty. Ltd.
11-13 Station Street (P.O. Box 163)
Mitcham, Vic. 3132   Tel. (03) 873 4411

AUSTRIA - AUTRICHE
OECD Publications and Information Centre,
4 Simrockstrasse,
5300 Bonn (Germany)   Tel. (0228) 21.60.45
Gerold & Co., Graben 31, Wien 1 Tel. 52.22.35

BELGIUM - BELGIQUE
Jean de Lannoy,
Avenue du Roi 202
B-1060 Bruxelles   Tel. (02) 538.51.69

CANADA
Renouf Publishing Company Ltd
1294 Algoma Road, Ottawa, Ont. K1B 3W8
Tel: (613) 741-4333
Stores:
61 rue Sparks St., Ottawa, Ont. K1P 5R1
Tel: (613) 238-8985
211 rue Yonge St., Toronto, Ont. M5B 1M4
Tel: (416) 363-3171
Federal Publications Inc.,
301-303 King St. W.,
Toronto, Ont. M5V 1J5   Tel. (416)581-1552
Les Éditions la Liberté inc.,
3020 Chemin Sainte-Foy,
Sainte-Foy, P.Q. G1X 3V6, Tel. (418)658-3763

DENMARK - DANEMARK
Munksgaard Export and Subscription Service
35, Nørre Søgade, DK-1370 København K
Tel. +45.1.12.85.70

FINLAND - FINLANDE
Akateeminen Kirjakauppa,
Keskuskatu 1, 00100 Helsinki 10   Tel. 0.12141

FRANCE
OCDE/OECD
Mail Orders/Commandes par correspondance :
2, rue André-Pascal,
75775 Paris Cedex 16   Tel. (1) 45.24.82.00
Bookshop/Librairie : 33, rue Octave-Feuillet
75016 Paris
Tel. (1) 45.24.81.67 or/ou (1) 45.24.81.81
Librairie de l'Université,
12a, rue Nazareth,
13602 Aix-en-Provence   Tel. 42.26.18.08

GERMANY - ALLEMAGNE
OECD Publications and Information Centre,
4 Simrockstrasse,
5300 Bonn   Tel. (0228) 21.60.45

GREECE - GRÈCE
Librairie Kauffmann,
28, rue du Stade, 105 64 Athens Tel. 322.21.60

HONG KONG
Government Information Services,
Publications (Sales) Office,
Information Services Department
No. 1, Battery Path, Central

ICELAND - ISLANDE
Snæbjörn Jónsson & Co., h.f.,
Hafnarstræti 4 & 9,
P.O.B. 1131 – Reykjavik
Tel. 13133/14281/11936

INDIA - INDE
Oxford Book and Stationery Co.,
Scindia House, New Delhi 110001
Tel. 331.5896/5308
17 Park St., Calcutta 700016   Tel. 240832

INDONESIA - INDONÉSIE
Pdii-Lipi, P.O. Box 3065/JKT.Jakarta
Tel. 583467

IRELAND - IRLANDE
TDC Publishers - Library Suppliers,
12 North Frederick Street, Dublin 1
Tel. 744835-749677

ITALY - ITALIE
Libreria Commissionaria Sansoni,
Via Benedetto Fortini 120/10,
Casella Post. 552
50125 Firenze   Tel. 055/645415
Via Bartolini 29, 20155 Milano   Tel. 365083
La diffusione delle pubblicazioni OCSE viene
assicurata dalle principali librerie ed anche da :
Editrice e Libreria Herder,
Piazza Montecitorio 120, 00186 Roma
Tel. 6794628
Libreria Hœpli,
Via Hœpli 5, 20121 Milano   Tel. 865446
Libreria Scientifica
Dott. Lucio de Biasio "Aeiou"
Via Meravigli 16, 20123 Milano   Tel. 807679

JAPAN - JAPON
OECD Publications and Information Centre,
Landic Akasaka Bldg., 2-3-4 Akasaka,
Minato-ku, Tokyo 107   Tel. 586.2016

KOREA - CORÉE
Kyobo Book Centre Co. Ltd.
P.O.Box: Kwang Hwa Moon 1658,
Seoul   Tel. (REP) 730.78.91

LEBANON - LIBAN
Documenta Scientifica/Redico,
Edison Building, Bliss St.,
P.O.B. 5641, Beirut   Tel. 354429-344425

**MALAYSIA/SINGAPORE -
MALAISIE/SINGAPOUR**
University of Malaya Co-operative Bookshop
Ltd.,
7 Lrg 51A/227A, Petaling Jaya
Malaysia   Tel. 7565000/7565425
Information Publications Pte Ltd
Pei-Fu Industrial Building,
24 New Industrial Road No. 02-06
Singapore 1953   Tel. 2831786, 2831798

NETHERLANDS - PAYS-BAS
SDU Uitgeverij
Christoffel Plantijnstraat 2
Postbus 20014
2500 EA's-Gravenhage   Tel. 070-789911
Voor bestellingen: Tel. 070-789880

NEW ZEALAND - NOUVELLE-ZÉLANDE
Government Printing Office Bookshops:
Auckland: Retail Bookshop, 25 Rutland Stseet,
Mail Orders, 85 Beach Road
Private Bag C.P.O.
Hamilton: Retail: Ward Street,
Mail Orders, P.O. Box 857
Wellington: Retail, Mulgrave Street, (Head
Office)
Cubacade World Trade Centre,
Mail Orders, Private Bag
Christchurch: Retail, 159 Hereford Street,
Mail Orders, Private Bag
Dunedin: Retail, Princes Street,
Mail Orders, P.O. Box 1104

NORWAY - NORVÈGE
Narvesen Info Center – NIC,
Bertrand Narvesens vei 2,
P.O.B. 6125 Etterstad, 0602 Oslo 6
Tel. (02) 67.83.10, (02) 68.40.20

PAKISTAN
Mirza Book Agency
65 Shahrah Quaid-E-Azam, Lahore 3 Tel. 66839

PHILIPPINES
I.J. Sagun Enterprises, Inc.
P.O. Box 4322 CPO Manila
Tel. 695-1946, 922-9495

PORTUGAL
Livraria Portugal, Rua do Carmo 70-74,
1117 Lisboa Codex   Tel. 360582/3

**SINGAPORE/MALAYSIA -
SINGAPOUR/MALAISIE**
See "Malaysia/Singapor". Voir
«Malaisie/Singapour»

SPAIN - ESPAGNE
Mundi-Prensa Libros, S.A.,
Castelló 37, Apartado 1223, Madrid-28001
Tel. 431.33.99
Libreria Bosch, Ronda Universidad 11,
Barcelona 7   Tel. 317.53.08/317.53.58

SWEDEN - SUÈDE
AB CE Fritzes Kungl. Hovbokhandel,
Box 16356, S 103 27 STH,
Regeringsgatan 12,
DS Stockholm   Tel. (08) 23.89.00
Subscription Agency/Abonnements:
Wennergren-Williams AB,
Box 30004, S104 25 Stockholm Tel. (08)54.12.00

SWITZERLAND - SUISSE
OECD Publications and Information Centre,
4 Simrockstrasse,
5300 Bonn (Germany)   Tel. (0228) 21.60.45
Librairie Payot,
6 rue Grenus, 1211 Genève 11
Tel. (022) 31.89.50
Maditec S.A.
Ch. des Palettes 4
1020 – Renens/Lausanne   Tel. (021) 635.08.65
United Nations Bookshop/Librairie des Nations-
Unies
Palais des Nations, 1211 – Geneva 10
Tel. 022-34-60-11 (ext. 48 72)

TAIWAN - FORMOSE
Good Faith Worldwide Int'l Co., Ltd.
9th floor, No. 118, Sec.2, Chung Hsiao E. Road
Taipei   Tel. 391.7396/391.7397

THAILAND - THAILANDE
Suksit Siam Co., Ltd., 1715 Rama IV Rd.,
Samyam Bangkok 5   Tel. 2511630
INDEX Book Promotion & Service Ltd.
59/6 Soi Lang Suan, Ploenchit Road
Patjumamwan, Bangkok 10500
Tel. 250-1919, 252-1066

TURKEY - TURQUIE
Kültur Yayinlari Is-Türk Ltd. Sti.
Atatürk Bulvari No: 191/Kat. 21
Kavaklidere/Ankara   Tel. 25.07.60
Dolmabahce Cad. No: 29
Besiktas/Istanbul   Tel. 160.71.88

UNITED KINGDOM - ROYAUME-UNI
H.M. Stationery Office,
Postal orders only: (01)873-8483
P.O.B. 276, London SW8 5DT
Telephone orders: (01) 873-9090, or
Personal callers:
49 High Holborn, London WC1V 6HB
Branches at: Belfast, Birmingham,
Bristol, Edinburgh, Manchester

UNITED STATES - ÉTATS-UNIS
OECD Publications and Information Centre,
2001 L Street, N.W., Suite 700,
Washington, D.C. 20036 - 4095
Tel. (202) 785.6323

VENEZUELA
Libreria del Este,
Avda F. Miranda 52, Aptdo. 60337,
Edificio Galipan, Caracas 106
Tel. 951.17.05/951.23.07/951.12.97

YUGOSLAVIA - YOUGOSLAVIE
Jugoslovenska Knjiga, Knez Mihajlova 2,
P.O.B. 36, Beograd   Tel. 621.992

Orders and inquiries from countries where
Distributors have not yet been appointed should be
sent to:
OECD, Publications Service, 2, rue André-Pascal,
75775 PARIS CEDEX 16.

Les commandes provenant de pays où l'OCDE n'a
pas encore désigné de distributeur doivent être
adressées à :
OCDE, Service des Publications. 2, rue André-
Pascal, 75775 PARIS CEDEX 16.

72380-1-1989